OPERATION AND MAINTENANCE OF LARGE INFRASTRUCTURE PROJECTS

PROCEEDINGS OF THE INTERNATIONAL SYMPOSIUM ON ADVANCES IN OPERATION
AND MAINTENANCE OF LARGE INFRASTRUCTURE PROJECTS
COPENHAGEN/DENMARK/10-13 MAY 1998

Operation and Maintenance of Large Infrastructure Projects

Edited by

Leif J. Vincentsen
A/S Storebælt, Copenhagen, Denmark

Jens Sandager Jensen
COWI, Lyngby, Denmark

Taylor & Francis
Taylor & Francis Group

LONDON AND NEW YORK

Organized by

The texts of the various papers in this volume were set individually by typists under the supervision of each of the authors concerned.

Published by Taylor & Francis
2 Park Square, Milton Park, Abingdon, Oxon, OX14 4RN
270 Madison Ave, New York NY 10016

Transferred to Digital Printing 2007

ISBN 90 5410 963 7

Publisher's Note
The publisher has gone to great lengths to ensure the quality of this
reprint but points out that some imperfections in the original
may be apparent

Operation and Maintenance of Large Infrastructure Projects, Vincentsen & Jensen (eds)
© 1998 Taylor & Francis, ISBN 90 5410 963 7

Table of contents

Operation and Maintenance of Large Infrastructure Projects, Vincentsen & Jensen (eds)
© 1998 Taylor & Francis, ISBN 90 5410 963 7

Preface

Major infrastructure links across water represent large investments. The structures and systems must be optimised to keep costs in control. Optimisation needs and the tendency to more slender and light structures imply that engineering disciplines like Bridge Aerodynamics and Ship Collision Analysis have an increasing impact on the overall design of the links. Also the attention to life cycle costs implies that Operation and Maintenance must be investigated and planned in parallel to the design and construction of the links.

The 1998 International Symposium aims at presenting state-of-the-art and future development trends within the three mentioned engineering disciplines. The background is the knowledge and experience gained in connection with the many recent major links built over water and the research carried out at universities and institutes.

The main focus areas within Operation and Maintenance of Large Infrastructure Projects are organisation, strategies and policies, traffic management and toll collection, O&M management as well as assessment and requalification of structures and equipment, which all are key parameters in the multidisciplinary optimisation effort after inauguration of the fixed links.

The proceedings contain papers by international experienced owners, operators and consultants, who have insight knowledge gained through practical experience within application of both basic and advanced state-of-the-art models, methods and tools.

Speakers have been invited to present their views of the present situation and development trends. We hope by this approach to attain a coherent presentation of the selected topics and a valuable set of proceedings. The proceedings are offset from the original manuscripts as provided by the authors. The editors are consequently not responsible for errors or misprints in the papers. Furthermore, the opinions of the authors are not necessarily those of the editors.

The organising committee and the editors convey their sincere thanks to the members of the scientific committee and to all the authors for their valuable contributions and their co-operation in the preparation of the proceedings.

Leif J.Vincentsen
Chairman of the Scientific Committee

Erik Yding Andersen
Chairman of the Organising Committee

Jens Sandager Jensen
Vice-chairman of the Scientific Committee

Operation and Maintenance of Large Infrastructure Projects, Vincentsen & Jensen (eds)
© 1998 Taylor & Francis, ISBN 90 5410 963 7

Organization

Symposium organized by
A/S Storebælt
Øresundskonsortiet
COWI Consulting Engineers and Planners AS

Main sponsor
The COWIfoundation

Co-sponsors
A/S Storebælt
Øresundskonsortiet
COWI Consulting Engineers and Planners AS
DTU Technical University of Denmark
DANHIT High Performance Computing and Networking

Co-sponsoring associations
Danish Group of IABSE
Permanent International Association of Navigation Congresses (PIANC)
International Bridge, Tunnel and Turnpike Association (IBTTA)
World Road Association (PIARC)

Organizing Committee
E.Y.Andersen, COWI (Chairman)
M.Clausen, MiaCon
F.Ennemark, Øresundskonsortiet
K.H.Ostenfeld, COWI
L.J.Vincentsen, A/S Storebælt
S.Esdahl, COWI (Secretary)

Scientific Committee
L.J.Vincentsen, A/S Storebælt, Denmark (Chairman)
J.Sandager Jensen, COWI, Denmark (Vice-chairman)
H.Ingvarsson, Swedish National Road Administration, Sweden
J.Lauridsen, Danish Road Directorate, Denmark
M.Michel Marec, Centre d'Etudes des Tunnels, France
H.C.Oud, Ministry of Transport, Public Works and Water Management, Netherlands

Strategies and policies for operation and maintenance

Operation and Maintenance of Large Infrastructure Projects, Vincentsen & Jensen (eds)
© 1998 Taylor & Francis, ISBN 90 5410 963 7

Management system for O&M based on the European Model for 'Business Excellence'

L.J.Vincentsen
A/S Storebælt (Sund and Bælt A/S), København, Denmark

ABSTRACT: Management systems are becoming more and more complex. It is no longer enough to focus on the financial results. Today, the customer and society require a commitment from the companies which often goes far beyond the activities in the individual company.

This is a fact which will become more apparent in the coming years. Therefore a broader management system like the Model for Business Excellence is a tool for future leaders.

This paper describes how the European Model for Business Excellence, together with the well-known management system standards for quality (ISO 9001), environment (ISO 14001) and occupational health and safety (BS 8800), can be implemented as an integrated management system for operation and maintenance.

1 FROM CONSTRUCTION TO OPERATION AND MAINTENANCE

When operation and maintenance begin, focus on problems during construction related to delays, overrun on costs and contractual disputes will fade out and attention will turn to the quality of the infrastructure.

The quality is defined in general at an early stage by the client, reworded by the consultants in drawings and technical specifications, transformed into the physical structures via the methods and materials used by the contractor and finally handed over to the operator together with instructions on operation and maintenance.

Quality of the infrastructure handed over after completion of construction will in fact not only be assessed by the client but to a high degree also by the users, the operator, the authorities and the public in general. They will deliver the real judgement.

The better the quality assurance is performed during design and construction, the more quality during operation will fulfil all expectations, provided that operation and maintenance are carried out as planned.

During the last decade, quality assurance has consequently occupied an increasing part of the design and construction management.

In the same decade risk management has become a general tool when designing the overall layout of infrastructure projects, for planning of construction and for setting up procedures to be used during

Figure 1. Typical management areas during construction.

operation. The risk analysis work is based on experience and statistics which date many years back in order to predict or forecast future events or accidents. If risks are identified and found too great (either because of major consequences or because of high probability) risk reducing measures are introduced. But are the results of the risk analyses still reliable when the real operation starts? This has to be monitored and taken care of by the O&M organisation.

During the design and construction of large infrastructure projects use of the expertise and the technology available worldwide is a decisive necessity as this is the only way for the technology to develop

Figure 2. The East Bridge. An infrastructure for which risk analyses have played an important role.

Figure 3. Simulation of navigation span.

further. Knowledge management is consequently an important factor in order to ensure progress and quality. Use of the best consultants for design and the best experts for design reviews is an insurance against bad quality. When moving at the edge of earlier experience, new effects will occur which must be predicted by the experts in order to achieve reliable solutions. This could be wind induced movements or ship collisions against bridge structures. If sufficient knowledge is not available tests or simulations will have to be performed. However, test facilities cannot fully reflect nature. In the operation phase the moment of truth will show up.

Environmental considerations play an increasing role in the design and construction of infrastructure projects. Today, often expenditure of 5 to 10% of the total costs is involved to improve the environmental impact of the structures. However, it is observed afterwards that sometimes the money could have been spent more efficiently on other areas, as nature often finds its own new balance in harmony with the structures. Mussels, eider ducks and algae blooms are often used as change indicators for the environmental conditions close to bridges. Some years after construction it is often seen that they are back to an even higher extent than before the start of construction. It is therefore important that the O&M organisation monitors the environmental conditions

and provides the necessary feed back to the experts on the developments.

The construction phase will result in a physical structure with a built-in quality which must stand the test during operation. The procedures handed over to the operator together with expectations and assumptions regarding how to use the infrastructure during operation are the first tools for proper operation. But they must not be a pretext for doing nothing. Constant improvements are required to keep pace with the changes in society, the environment, the technology and the traffic volume.

2 MANAGEMENT SYSTEM FOR THE OPERATION PHASE

Operating a large infrastructure system it is of vital importance to bring into focus the results from operation that create most value for the interested parties which are the owner, the customers, society and the employees (people).

It is important for the management not only to look at the economic results (the owner's perspective), but to a higher degree also to turn its eye towards the customer - the road user. With the increasing awareness of environment and safety and with the increasing social importance of effective infrastructures the results of operation and maintenance must also be seen in a social perspective. Last but not least, it is evident that good service and excellent results are created only through motivated and satisfied employees. Consequently the people's (employees) satisfaction must be another cornerstone of the management.

Excellent results are created via effective leadership, leading the way with clear objectives and strategies and continuously focusing on possible improvements based on measurement of facts and systematic evaluation of results obtained.

The resources forming the basis for good results are not only the employees but also external personnel, technology, material and equipment and information available.

The European Model for Business Excellence is a model for leadership which focuses on the essential enablers and which determinedly works to create a balanced success for all interested parties. This value-based management model has been taken into use in more than 30 countries and forms the basis for the Danish quality award and the European quality award for excellent business. It systematizes the afore-mentioned enablers and results.

Organisations using the model are not looking for short-term economic gains, but focus to a higher degree on the satisfaction of customers and people and on the impact on society. The conviction is that in the long perspective this gives the desired economic results.

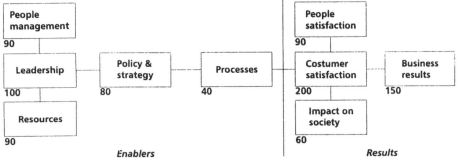

Figure 4. Model for Business Excellence.

The management model is not only to be used by the operator of an infrastructure system, but is also of value to the consulting engineers during design of the infrastructure as they create essential parts of the basis for operation.

The management model implies that within each result area clear objectives - which the interested parties of course find relevant and essential - are defined with operational measurement points, from which objective data can be collected and trends evaluated.

A performance scorecard based on this will show the factors which in the long term create value for the organisation, contrary to economic accounts which only measure costs and the economic value created.

The balanced performance scorecard must at all times be tailored to the situation for the individual organisation. It is a continuing process of improvement.

For the O&M organisation of a large infrastructure system it is important at the strategic level to identify six to eight key figures within each group of interested parties, which can be regarded as the critical success factors and which look sufficiently ahead (3-4 years).

These key figures are compared every month/year in order to follow the trend and to implement improvements in due course where necessary.

An organisation's business results presented on a performance scorecard can be compared with other organisations' scorecards. It becomes more and more common to perform national and international benchmarking among organisations in order to find how the performance is evaluated and classified compared to other organisations.

One could ask what the difference is between the Total Quality Management (as described in the Model for Business Excellence) and Quality Management based on ISO 9001?

The latter is often based on management of the processes in the organisation through preventive and corrective actions and through assessment from quality audits performed. Thus quality management based on ISO focuses more on the enablers than on the results.

Characteristics of the management model based on Business Excellence:
- is a holistic concept
- focus on customer satisfaction
- creates a culture of innovation
- is based on facts
- place people (employees) in the centre
- requires visible and committed leadership
- focus on internal as well as external quality
- is effective for communication
- requires vision and broad view

Figure 5. Characteristics of the management model based on Business Excellence.

Even though an organisation has an effective certified quality management system based on ISO 9001 it may be far from obtaining a high score according to the business excellence model.

The contents of the environmental management system (ie based on ISO 14001) and the quality management system (ie based on ISO 9001) are shared with regard to 60-70%. Consequently it is natural to integrate these two types of system in one and to perform audits on both systems at the same time. Today, environmental management covers both the external and internal environment (working environment).

Environmental management is introduced as a result of requirements from authorities, customers, investors, based on ethical motives or marketing advantages. The organisation often decides - but is not required - to set up a so-called "Green account" for publication. This "Green account" describes the environmental impact of the business's activities, consumption of energy, water and raw materials and the nature and quantity of waste and polluting substances.

It can be expected that not only financial key figures play a role for future lender decisions but so do the figures from the "Green account", the so-called environmental indicators.

Management of the external environment is normally based on an environmental examination of all activities going on in the organisation, on which

5

occasion a balance sheet is drawn up for handling of the environmental aspects.

Management of the working environment is normally based on workplace examinations for the individual employee. Such an examination and the resulting corrective actions contribute to the satisfaction of the employees.

Integration of quality and environmental management, together with the Business Excellence model, is the obvious solution for a management system for the future. It is, however, important that such an integrated system does not become a specialist discipline for a staff function in the company.

The management system must be deeply rooted in the whole organisation so that all employees, external resources and suppliers are involved in creating the relevant values for the interested parties.

It is one of the biggest challenges for today's leader to change attitudes and to get everybody motivated on a daily basis to focus on balanced quality and to be environmentally aware. This is the case when buying materials and equipment, when signing service contracts and when performing daily work routines.

3 IMPLEMENTATION OF THE MODEL FOR BUSINESS EXCELLENCE AT THE GREAT BELT PROJECT

This section describes experience gained during implementation of a management system at the Great Belt Project based on the model for Business Excellence and integrated with systems for quality, environment and safety.

Facts about the Great Belt Project:
- a 6.6 km low-level bridge for road and rail traffic
- a 6.8 km suspension bridge with the second longest span in the world (1624 m)
- an 8 km bored railway tunnel
- extensive reclamation and landworks
- daily road traffic volume in 2001: 14,500 passenger cars and 2,700 lorries
- first toll collection system in Denmark
- highest toll revenue per year at one toll plaza in the world
- first VTS centre for supervision of ship traffic in Denmark

---

Figure 6. Facts about the Great Belt project.

In November 1995 a pilot project was carried out in the technical department with the purpose of testing the Business Excellence model on a smaller scale and to familiarize the staff with the TQM concept. In this way some of the employees were trained in the implementation of the model in the entire organisation.

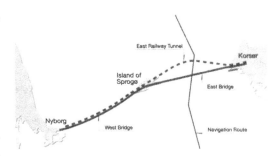

Figure 7. Map of the Great Belt link.

The pilot project had a duration of almost 1 year and consisted primarily of a self-assessment of the department's five enablers, see figure 4. The result was several notes describing the strong and weak points of the department, giving input for improvements. This result was an excellent feedback to the leader of the technical department, feedback which was difficult to obtain via regular monthly meetings and the yearly appraisal interview.

The main contents of the model were presented to the management at the end of an internal 2-day seminar in June 1996. The topics of the seminar were objectives for the organisation during operation and outsourcing as a business strategy.

As a result of the seminar four of the managers established a working group in September 1996 with the purpose to formulate a proposal for the overall policy and strategy for the organisation in the operation phase.

This strategy was approved by the managing directors in November 1996.

Preparations for operation and maintenance continued in the spring of 1997, mainly of the technical area (with the setting up of technical procedures) but also with preparation for the organisational set-up.

At that time it was decided to integrate the organisation with the organisation for the Landworks of the Øresund Link. - Another large infrastructure project under construction in Denmark in these years.

Furthermore, in April 1997 a consultant was engaged in order to assist the organisation in the implementation of the overall management model for Business Excellence combined with management systems for quality, environment and safety.

Contrary to traditional management systems this system was planned to
- cover all business activities in the organisation
- cover all departments and functions
- be based on cross-functional as well as functions-specific procedures
- cover all requirements in the standards for quality, environment and health and safety.
- describe activities within all 9 areas of the Business Excellence model.

In order to secure a proper and broad commitment to the management principles a steering committee was established consisting of the 4 managers, the consultant and the TQM leader of the technical department.

The structure of the integrated management model can be illustrated by the following figure:

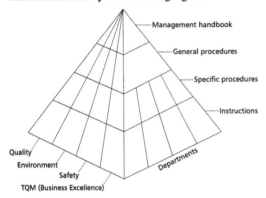

Figure 8. The structure of integrated management.

The model has four levels. The upper level contains the management handbook for the organisation describing the company, the overall policy and the strategy for the organisation and the activities within the individual departments.

The second level consists of 20 general procedures which by use of flow diagrams describe the guidelines for the general or cross-functional procedures in the organisation.

The 20 general procedures are listed in the following figure:

GP 01 Management and organisation
GP 02 Management self-assessment
GP 03 Meeting structure
GP 04 Reporting
GP 05 Management system structure
GP 06 Marketing, sale and external communication
GP 07 Development of services
GP 08 Control of the management system
GP 09 Documents and data management
GP 10 Purchasing and suppliers
GP 11 Management of deviations
GP 12 Corrective and preventive actions
GP 13 Audits
GP 14 People management
GP 15 Statistics
GP 16 Environment
GP 17 Safety and health
GP 18 Financial control
GP 19 Information technology
GP 20 Insurance

Figure 9: General procedures.

Levels 1 and 2 of the management system were established in a first edition from May to September 1997.

Levels 3 and 4 are department-specific procedures. As for the technical department these procedures have been under development since early 1995 internally and from August 1996 with the assistance of external consultants. They consist of about 70 technical procedures and almost 100 technical instructions.

The structure of levels 3 and 4 is closely related to the structure of the planned organisation in order to obtain a user-friendly system. All procedures and instructions are planned to be available on the intra-network of the organisation so that use becomes simple and the users are guided from procedure to instructions and further to other relevant documents and are always sure to look up valid editions.

Parallel with the development of the management model the organisation for operation and maintenance was described. An overall strategy was set up for integration of the organisation for O&M of the Great Belt Project and the Danish Landworks for the Øresund Link. These two projects would both be in operation in 1997 and 1998.

Furthermore, it was agreed to coordinate O&M activities with the Øresund coast-to-coast link owned on a 50/50 basis by a Danish and a Swedish partner.

The model for the organisation was presented in June 1997 and the employees were engaged in November 1997 to start the operation of the Great Belt road link on 14 June 1998 in the joint operations organisation for the two projects, named Sund & Bælt A/S.

The organisation is small and consists of only 36 employees in total as the strategy from the beginning is to outsource as much as possible.

This is to ensure flexibility during the first years of operation as it is easier to build up than to reduce the size of an organisation.

After having engaged the employees, two management seminars were held in December 1997 and January 1998 with the purpose to rewrite and detail the overall policy and strategy and to plan the implementation of the management system in more

Great Belt	
Rail link	1 April 1997
Road link	14 June 1998
Øresund Landworks	
Road link	26 September 1997
Rail link	27 September 1998
Øresund coast-to-coast link	
Road and rail link	Summer 2000

Figure 10. Time schedule for inaugurations.

Figure 11. Organisation of Sund & Bælt A/S.

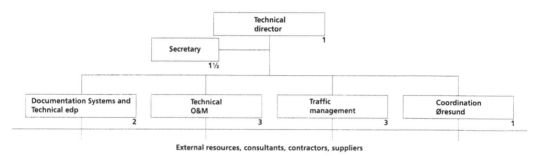

Figure 12. Organisation of the technical department.

detail. A kick-off meeting with all employees was held in January 1998 and a team-building seminar for employees in March 1998 with the purpose of getting together and to learn how to change the focus from construction to customer service.

4 OBJECTIVES IN THE MANAGEMENT SYSTEM FOR AN INFRASTRUCTURE ORGANISATION

An O&M organisation for a large infrastructure project has many different tasks. Some of them are monitoring and maintenance of structures and installations, traffic management, toll collection, sale of subscriptions and paying back construction loans.

The great variety of tasks requires a clear business strategy. For the Sund & Bælt A/S organisation the strategy contains the following overall objectives:

Overall objectives for Sund & Bælt A/S:
- to make Sund & Bælt A/S an effective, modern service organisation with a businesslike approach
- to supply a high level of service and to run the organisation with high priority on satisfaction of people
- to work actively to increase the utility value of the infrastructure to society
- to offer a flexible, reliable, available and competitive transport facility with a large capacity which can be easily adjusted to the customers' needs, the development in society and the economic objectives
- to work for close supervision of and improvements in the environmental and safety areas within the activities performed by the organisation

Figure 13. Business strategy for Sund & Bælt A/S.

These overall objectives must be transferred into operational and concrete objectives within each department to cover the four result areas, see Figure 4, the Business Excellence model.

Examples of such operational measurement tools are given in figure 14.

People (employee) satisfaction:
- measurements of the quality of the manager's leadership
- yearly appraisal interviews and follow-up
- yearly investigation of employee satisfaction
- extent of communication with the employees on policies and results
- employees' turnover and absenteeism rate
- statistics on training and education
- number of cases for the joint council

Customer satisfaction:
- yearly investigation of customer satisfaction
- meetings with customer panels
- measurement of waiting time and tailbacks
- measurement of availability of traffic lanes due to bad weather conditions, accidents, work, heavy transport, etc.
- analysis of complaints from customers

Impact on society:
- statistics of traffic volume and composition
- yearly investigation of company image
- measurement of energy consumption, waste volume
- measurement of traffic accidents
- statistics of industrial accidents
- registration of use of environmentally correct products
- participation in committees on social issues
- visualize social impact at lectures and in articles

Company business results:
- annual reports based on holistic principles
- sales volume per customer fraction, type of subscription, means of payment, type of lane, etc.
- turnover and profit
- market shares
- payback time on loans

Figure 14. Examples of measurement tools within the four result areas.

5 EXAMPLES OF MEASUREMENT POINTS IN DAILY OPERATION

Many activities regarding operation of an infra-structure system are often outsourced and consequently clear agreements on the objectives and the quality of the service have to be made. In the following an example of quality measurement points is described.

Manning of toll collection booths

The people manning the toll collection booths for an infrastructure system are the face of the entire company. Consequently they must be able to deliver a friendly, obliging and courteous service to the customer which reflects the image the company wants to maintain.

It is required that they possess good language and calculation skills, are familiar with the use of computers, have passed a service education, have a well-groomed appearance and have no evidence of previous convictions.

Furthermore it is required that they spend a minimum number of active hours in the toll booth per month in order to keep up skills and that each year they participate in a certain number of hours of supplementary training.

During operation the quality of the service given by toll attendants is measured in different ways:
- analysis of complaints from customers
- questionnaire investigations of customer satisfaction
- anonymous sample buys
- measurement of service time
- examination of procedure knowledge.

6. CLOSING REMARKS

Management systems are becoming more and more complex. It is no longer enough to focus on the financial results of the company. Today, the customer and society require a commitment from the companies which often goes far beyond the activities in the individual company.

This is a fact which will become more apparent in the coming years. Therefore a broader management system like the Model for Business Excellence is a tool for future leaders.

Operation and Maintenance of Large Infrastructure Projects, Vincentsen & Jensen (eds)
© *1998 Taylor & Francis, ISBN 90 5410 963 7*

Infrastructure management systems applied to bridges

Bojidar S. Yanev
Department of Transportation & Columbia University, New York City, N.Y., USA

ABSTRACT: The article attempts to identify the fundamental tasks currently united under the term Bridge Management and the key developments of the transportation infrastructure in the United States which have marked the bridge management process over the recent decades. The New York City bridge management, past and current, is reviewed in some detail for the purpose of discerning trends and conclusions on the subject.

1 INTRODUCTION:
BRIDGE MANAGEMENT

Infrastructure management is currently considered as an imperative for maintaining and improving the quality of life worldwide. Transportation networks and bridges in particular have played a major part in recognizing the importance of the infrastructure for all aspects of industrial and social development. The experience of the United States in this domain is instructive. There are 280,898 state and interstate bridges and 309,142 local ones in the United States. 25% of the former and 36% of the latter or 31% of the total 590,040 are considered substandard (Better Roads, Nov. 1997). The Federal Highway Administration (FHWA) considers approximately 104,000 bridges as "Structurally Deficient". In recognition of this state of the national infrastructure, typical of the last decades, Congress passed the Intermodal Surface Transportation Efficiency Act (ISTEA) in 1991. As a result, bridge management became a mandated or recommended part of the activities of all State Transportation Departments. Over a period of several years the bridge management programs PONTIS and BRIDGET were developed by FHWA and adopted for implementation by a majority of the States. A number of other States, including New York, Pennsylvania and Indiana developed their own bridge management programs. Since 1994 it is also required to include life-cycle cost considerations in the design of new bridges. A current FHWA project

is developing an algorithm and software for life-cycle costs analysis. Yet most of the bridge management decisions are subject to important constraints beyond the immediate sphere of bridges. Consequently bridge management is not advancing at the rate with which various bridge management systems are developing.

A brief review of current trends in bridge management and bridge management systems follows. Detailed examples are presented regarding the 847 bridges managed by the Department of Transportation, City of New York.

2 THE CONSTRAINTS

It is increasingly recognized that a systematic approach to the management of both national and local infrastructure networks requires a commitment on all levels. Figure 1 illustrates this point with a statement by the President of the United States.

In a large country governed by democratic principles, however, such a commitment does not lead directly to the adoption and implementation of a national management policy.

Some of the problems are summarized by the (then) Chief Administrator of FHWA and now Transportation Secretary Rodney E. Slater in a follow-up to the letter of Figure 1, dated October 21, 1993. In it Secretary Slater states: "...Needs typically exceed the means available to address them. At the national level, ISTEA has allowed us to set new funding records for highway and bridge projects, but still, needs far exceed the available funding. Resources, therefore, must be distributed as equitably as possible....Clearly, if funds were unlimited, we would do more. However, we are proud of our role in helping New York City preserve its major bridges as important transportation links and as a valuable historic legacy for generations to come."

The main infrastructure management difficulties may be stated as follows:

Funding is always limited.

This is true by definition, since demand always exceeds the means. Consequently, optimization of the expenditures is required, subject to constraints and priorities. The funding constraint, while an obstacle to bridge maintenance is the reason for the development of management systems.

Management priorities are not similar on different government levels.

The Federal Government stirs vast resources towards the upgrading of the infrastructure, recommends and supervises their expenditure. Yet the Federal authority has its limitations. Within the local constraints States, municipalities and communities have executive power on the local level. As a result interstate and local infrastructure conditions are strikingly different. It was pointed out above that, while the interstate bridge network is estimated to have 25% deficient bridges, the local one shows 36% deficiency. The differences cannot be instantly resolved. ISTEA *mandated* the implementation of Bridge Management Systems (BMS) by the States in 1991, but, subject to pressure form the U. S. Senate, merely *recommended* the same in 1996. Eligibility for Federal funding is considered as an incentive for implementing BMS by the States.

Long term bridge management is not independent within the economy.

Bridge management is part of *transportation management* which, in turn, is part of *infrastructure management.* Infrastructure management is an important part of the entire economic structure. The

October 6, 1993

Mr. Bojidar S. Yanev
Assistant Commissioner
Bridge Inspection/Research
   and Development
New York Department of Transportation
Fourth Floor
2 Rector Street
New York, New York 10006

Dear Mr. Yanev:

Thank you for your thoughtful letter regarding New York City's infrastructure. I agree with you that America must address the problems of its vast network of bridges and highways if we are to remain a strong nation in the next century, and I have forwarded your letter to the Department of Transportation for futher review. I will keep your ideas in mind as I face the great challenges ahead.

Sincerely,

Bill Clinton

Figure 1. Letter of United States President Clinton.

relationship between infrastructure condition and economic growth is the subject of considerable attention. It is generally accepted that the economy drives infrastructure development, but a reverse causality with the infrastructure actively influencing the economy is also considered. With the general subject of infrastructure management under various stages of development, the funding available to most bridges, an essential constraint of the management algorithm, remains indeterminate.

3 THE STRATEGIES

The source of funding, dedicated to the infrastructure facilities, such as the road and bridge networks, has become a fundamental issue of infrastructure management. By the same token, the management of structures where dedicated funding is in place, such as the toll bridges, is relatively better developed and free of the generally observed deficiencies. Most structures on the Federal and local transportation network are not in this category. It is their management that is currently under intensified scrutiny.

This is aptly demonstrated by a comparison between the management of bridges, comparable in size and function, but with different sources of funding. Yanev (1994) draws such a comparison between the George Washington and the Williamsburg bridges, both in New York City. The former is a toll bridge managed by the Port Authority of New York and New Jersey, currently earning approximately $250 million annually. It was built 65 years ago and is considered to be in excellent condition . The latter is managed by the City of New York and is currently undergoing a rehabilitation at a cost expected to reach $700 million in Federal, State and local funds. The maintenance of the George Washington (currently costing about $20 million/year) is viewed as a protection of this highly profitable facility. The rehabilitation of the Williamsburg and its future maintenance are a burden to the taxpayers, and thus occupy a highly unpopular position among budget priorities.

The above comparison exemplifies the two fundamentally different bridge management strategies as follows:

Minimize immediate (first and current) costs.
Minimize life-cycle costs.

During the Pre- and Post-World War II decades the fundamental management policy nationwide was defined by a vast Federal investment in new construction. The pattern was retained and the idea behind many bridge management decisions was to obtain maximum service from a structure at minimum cost (as for maintenance) and then replace it. On the local level this strategy had an additional attraction since bridge reconstruction was Federally funded, while maintenance was funded by local taxes. Only in 1993 did Executive Order 12893 (FHWA) "Principles of Federal Infrastructure Investment" mandate that benefits and costs of infrastructure investment be measured and appropriately discounted over the full life cycle of the project. The current NCHRP Project 12-43 is charged to develop a methodology for Bridge life-Cycle Cost Analysis (BLCCA) for transportation agencies.

4 THE STATE OF THE ART

4.1 Background

The weaknesses of the minimized first cost strategy began to surface as the bridges progressively aged.

In a toll bridge the decline in service would immediately translate into reduced income and thus become a management priority. In a bridge with no direct connection between user satisfaction and structural condition, the first warning of a bridge management lapse could be structural failure. The collapse of the Silver Bridge over the Ohio River at Point Pleasant on December 15, 1967 alerted the United States to the imperative of introducing bridge management considerations nationwide. The decisions taken in the aftermath of that event can be viewed as the origin of the contemporary systematic bridge management efforts.

A number of FHWA and Transportation Research Board (TRB) publications contain the guidelines on bridge management in the United States. Primary among them are the Guide for Bridge Inventory and Appraisal, FHWA (1988), the Bridge Management Systems Report, NCHRP 300 (1987) and PONTIS, TRB Circular 423 (1994) , TRB Records 1183 and 1184 (1988), 1290 (1991), 1490 (1995). The Bridge Management Report by OECD (1992) summarized a great deal of Bridge management knowledge from all the member countries. Since 1995 PONTIS and BRIDGIT, the Federally developed BMS, are fully operational and adopted by over 40 States. The State of Pennsylvania, followed by New York, Indiana and others developed their own BMS.

4.2 Database

Both bridge management reports NCHRP 300 and OECD 1992 show schematic representations of the conceptual BMS. A fundamental feature of both is the database. Beginning in 1971 and following the guidelines of the Inventory Manual, the U. S. National bridge database was developed to include a full bridge inventory and annual or at least biennial bridge inspection and condition rating reports. As a result, the bridge database of the United States currently contains a complete inventory and a 20 year history of condition ratings for all bridge components. With the growing experience and quantity of bridge inspection, however, as pointed by Yanev (1996), certain deficiencies of the bridge condition data have emerged. Primary among them are:

- Subjectivity of condition ratings;
- Lack of quantified deterioration assessment;
- Lack of historic data on repair and maintenance work (both past and current).

Subject to the above deficiencies is the next step in the bridge management process, namely the decision making or analytical one.

4.3 Decision making

The second key function of the bridge management system, also prominently featured by all bridge management reports, including the two quoted above, is the decision making module. Bridge management decision making performs two fundamental tasks:

- Needs assessment (current and future);
- Funding optimization.

The attention of U.S. bridge managers in the early 1970's was directed, appropriately, to mitigating imminent hazards. Thus, the first decades of bridge inspections were spent in identifying numerous potentially hazardous conditions, leading to an exponential escalation in numbers and a substantiald evaluation in gravity of the latter term (Yanev 1994). With this period presently in the past, attention is, once again, directed towards performing the two task above over the useful life of the bridges. It is hoped that the latest versions of PONTIS, particularly after including the life-cycle costing analysis will be performing some of them.

4.3.1 Needs Assessment

The needs under consideration are usually those of the existing structures and not those of the traffic network. Thus, both the immediate and forecast bridge needs are the subject of structural engineering analysis. Attempts to estimate current bridge needs on the basis of existing inspection reports have been conducted and described by numerous bridge owners in the United States and Europe. In the case of a limited number of bridges (say under 300) the assessment usually addresses all of them. For larger numbers constructing a meaningful sample becomes necessary. Interesting work in the latter area was conducted and reported by SETRA and LCPC in France (Robichon & Binet & B. Godart, 1995).

Forecasting of bridge needs is based on models of deterioration for the bridges and their components. One frequently arising question is whether the bridge condition and deterioration issues should be addressed in a deterministic or in a probabilistic manner. The probabilistic approach, particularly with the application of Markov chains, has been adopted by most algorithms forecasting bridge conditions. The position taken herein is that the probabilistic and the deterministic approaches are essentially inseparable in this area and that, regardless of the point of departure for a particular model, the outcome, if correct, should not differ significantly.

4.3.2 Funding Optimization

This is the essential management function which determines the outcome of the entire process. The decisions taken on this level require expertise in economics as well as engineering. As stated on the onset, a coordination with other fields of the economy is essential. In recent decades such decisions have tended to remain firmly in the competence of the elective executive branch of the United States government with economists providing expertise and engineers advising the economists, e.g. twice removed from direct management power. It has been argued that distancing the engineers from the most important decisions on bridge management directly accounts for the decline in bridge conditions during the same period. This is, of course, the reverse way of observing that, with a sound infrastructure in place, the economy, even though also sound, chooses to ignore maintenance and maximize short term profit. Under these conditions optimizing any transportation infrastructure funding requires expertise far exceeding the structural engineering field. Hence, it can be concluded that the management process itself requires a certain refinement. To play a dominant role in this process, engineers will need advanced understanding of economics and the skill of imposing their position. Certain weaknesses of the economic analysis of infrastructure and, specifically, bridge needs have already been identified. For instance, (Yanev 1994), present worth analysis misleadingly devalues the benefits from the service of a structure beyond a certain time-span, depending on the discount rate. This may have led to the tendency of selecting lowest first cost structural designs regardless of their life-cycle prospects and, consequently, to a bridge life-span currently declining to approximately 30 years. On the other hand, even though engineers would prefer maintaining all of their bridges in the best condition possible, rigorous economic analysis seems to suggest that, under the inevitable funding constraints, the optimal maintenance level is always less than the maximum one, resulting in an "average" rather than "very good" bridge condition. The most widespread bridge management strategy is a more or less elaborate "triage", such that the structures in the worst condition are kept safe at minimum cost until they can be replaced, new structures are maintained to the extent that funding allows, and the structures in between are the subject of detailed needs analysis. Some BMS already in use provide recommendations on the optimal strategy for individual structures, and even structural networks, but all systems eventually call for an engineering decision. The thrust of current bridge management research is to demonstrate that intensifying the maintenance of good structures, as well as investing in better quality initial construction is economically preferable over the structural life cycle.

5 BRIDGE MANAGEMENT IN NEW YORK CITY

The above general management outlines gain significance when applied to the case of the New York City bridges.

5.1 History

The major bridge construction occurred at the turn of the 20th Century when the City of New York was established (Jan.1, 1898). The powerful Bridge Department reported 45 bridges in 1912. Their estimated construction cost was $71 million, annual maintenance was $1,280,490 or 1.8% of the above. The land acquisition cost was roughly $40 million. The oldest bridge was from 1693, but the average bridge age was 15 years. They were in good condition. Tolls were abolished in 1911. Large expansion was anticipated.

The current bridge ownership and responsibility took shape in the 1930's and 40's with the establishment of the Port and MTA Bridge and Tunnel Authorities. Some of the City's most famous and largest bridges, such as the Verrazano (1964), the George Washington (1931), Triboro, Whitestone (1937), Throg's Neck and Bayonne are managed by the two Authorities who charge tolls for crossing. The Hell Gate (1912) is a railroad bridge operated by Amtrak. Approximately 600 mostly post World War II highway bridges are the responsibility of New York State as part of the

Interstate network. Several thousand spans (mostly steel structures) support elevated sections of the City subway system.

847 bridges with approximately 4500 spans are managed by the City of New York. The Brooklyn (1884), Williamsburg (1903), Manhattan (1908) and Queensboro (1912) bridges over the East River are included. 85% of the structures are steel. In the post-WWII years the City Bridge Department gradually shrank to a small section of the Department of Transportation and the East River bridges were temporarily transferred to State responsibility. The average age of the New York State, the Port and MTA Authority bridges varies from 36 to 45 years. The average age of the 847 New York City bridges is roughly 67 years without adjustment for bridge rehabilitations.

Maintenance of the toll bridges remained at roughly the original level (1.8% of construction cost), while the City kept the total bridge maintenance expenditure approximately constant (after discounting) even as the bridge number increased by a factor of 50. Structural conditions declined and a span of the West Side Highway, built in 1928, collapsed in January of 1973. A diagonal stay broke on the Brooklyn Bridge, killing one pedestrian in 1980, alerting engineers to the fact that the entire suspension system needs replacement. In 1987 the Williamsburg bridge was closed due to structural deterioration and its replacement was seriously considered.

5.2 Organizational Structure

As the above developments reflect the National trend, so does the ensuing reaction. The Bridge Bureau was reestablished at the Department of Transportation, the responsibility for the East River bridges was resumed and a 10 year capital program for bridge rehabilitation addressing approximately 350 bridges was funded from Federal and local sources. A component rehabilitation program was also initiated in 1991.

The structure of the Bridge Bureau, fairly constant since that time, is illustrated on Figure 2. Currently called Division of Bridges, it includes Construction, Design, Maintenance and Inspection/Management Units. Bridge Construction primarily supervises work by contractors, Design is divided into sections for consultant supervision and in-house design, Maintenance conducts work both in-house and through contracts, Inspection/ Management conducts work in-house but also receives inspection reports for most of the City bridges from consultants working under contract for New York State.

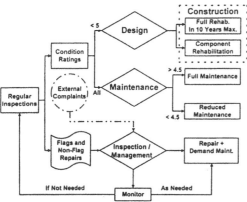

Figure 2. Bridge Division, Department of Transportation, New York City.

The New York State Bridge Management system is operational for most of the New York City bridges. This extends primarily to the database, consisting of bridge inventory and condition ratings. The strategy optimization module is still of limited use and the triage type of decision is applied as shown. Figure 2 refers to overall bridge condition ratings according to the New York State system, such that 7 indicates new structure, 5 - good condition, 3 - not functioning as designed and 1 - failed. The even numbers are for intermediate conditions. The overall bridge condition is a product of condition ratings for all components in all spans. This type of condition assessment has been repeatedly questioned (Yanev 1994) but it is unquestionably useful as an illustration of general trends. Condition ratings and hazard condition reports (flags) determine the plans for bridge maintenance, repair and reconstruction. New York State Department of Transportation states that the current bridge condition rating was weighted towards the assessment of bridge reconstruction needs. A different weighted average will be introduced for the assessment of maintenance needs.

5.3 Maintenance Optimization

The major New York City bridge management activities and their respective annual costs are

shown on Figure 3. It is indicated that over at least 10 years the overall average condition rating for the City bridges has remained nearly constant at approximately 4.5. This has been achieved at the annual costs shown in the figure.

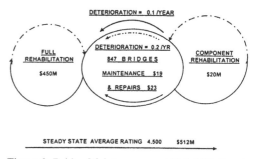

Figure 3. Bridge Maintenance and Rehabilitation in New York City.

Over the last 10 years the understanding of the rates of bridge deterioration, the maintenance needs and the possibilities for a more effective fund allocation has evolved. The following are main stages in this process.

5.3.1 The Preventive Maintenance Report 1989

The Preventive Maintenance Report for the Bridges of New York City, (Columbia University 1989) estimated "full maintenance" tasks for the City bridges at an annual cost of $56 million. This cost was considered equal to approximately 0.5% of the total bridge replacement cost, compared to maintenance levels estimated at 2.5% of replacement cost in Japan and 1.0% to 1.5% in Western Europe. The Report argued that the optimal maintenance could extend the bridge useful life up to 120 years, while, without maintenance bridges could fail after 20 years of use. Thus, if the total replacement cost of the bridges were estimated at $12 billion, the "full maintenance" strategy would cost $ (12,000 / 120 + 56), or $156 million annually, while the "no maintenance" annual cost would amount to $12,000/20, or $600 million. It was estimated that New York City bridge maintenance in the post-WWII years had slipped to 0.05% of the replacement cost, thus obtaining about 30 years of useful life per bridge at approximately $400 million annually. All dollars of the above paragraph are 1989. Detailed maintenance recommendations included a full paint cycle of 8 years, regular washing, oiling, etc.

The City accepted the findings of the Reprot and decided to fund full maintenance for bridges in good condition and reduced + "demand" maintenance (e.g. hazard mitigation) for bridges awaiting reconstruction.

5.3.2 New York City Bridge Categories 1990, 1997

It had been the City's long-standing practice to report of bridge conditions in terms of Poor, Fair, Good and Very Good. Concurrent with the decision to fully maintain Good and Very Good bridges was the decision to include the Poor and Fair bridges in the ten-year capital reconstruction plan. The New York City Department of Transportation Annual Bridge Report, 1990 for the first time related the above four adjectives to the New York State numerical ratings as shown in Table 1.

Table 1. New York City bridge conditions in 1990

Category	Rating	Bridges	%	Spans	%
Poor	1.0 - 3.0	53	6	1,205	24
Fair	3.01- 4.5	432	50	2,331	47
Good	4.51- 6.0	337	38	1,181	24
Very Good	6.01- 7.0	53	6	253	5
Not Rated		2	0	2	0
Total		877	100	4,972	100

That classification was considerably less optimistic than previous reports issued independently of the numerical condition ratings. For the first time it was recognized that Poor and Fair bridge conditions may not be eliminated by the year 2000.

Table 2. New York City bridge conditions in 1996

Category	Rating	Bridges	%	Spans	%
Poor	1.0 - 3.0	48	6	594	13
Fair	3.01 - 4.99	524	62	3,189	68
Good	5.0 - 6.0	148	17	658	14
Very Good	6.01 - 7.0	59	7	155	3
Not Rated	-	68	8	78	2
Total		847	100	4,674	100

In 1997 it was decided that the numerical rating of 5 should be the qualifying boundary between Fair and Good bridges for rehabilitation programs, while 4.5 should remain the limit for inclusion in Full Maintenance. This more stringent approach, too radical for 1990, was already acceptable in 1997. The resulting distribution is quoted in Table 2 from the New York City Department of Transportation Annual Bridge Report 1996:

5.3.3 Bridge Deterioration Rates

The first broad statement describing bridge deterioration as a function of maintenance was quoted above from the Preventive Maintenance Report. Subsequent studies (Yanev & Chen, 1993) indicated that the average bridge and bridge component annual rate of deterioration appears to be approximately 0.1 of a rating point, resulting in roughly 50 to 60 years of useful life. Peculiarities of the deterioration curve behavior were noticed, in agreement with other publications, suggesting that the rate of decline decelerates as it descends below the rating of 4. and never reaches 1. Both the 0.1 rate and the asymptotic approach of the rating of 3 suggested that the obtained curves show the effect of undocumented rehabilitation and repair work and do not represent no-maintenance rates of deterioration.

Other publications (Llanos & Yanev 1991), (Yanev 1997) sought to exclude or at least minimize the effect of maintenance and repair by selecting only the lowest, rather than the average, ratings for bridges and/or components of any age. The results clearly showed deterioration rates of 0.2 of a point for bridges, primary members and decks and much faster deterioration rates for joints, scuppers and wearing surfaces. Typical "worst case" deterioration curves are shown on Figure 4 along with the "average" bridge deterioration rate.

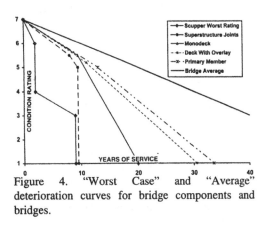

Figure 4. "Worst Case" and "Average" deterioration curves for bridge components and bridges.

The results are obtained and treated deterministically, however, a probabilistic approach should not produce a significantly different outcome. The findings are consistent with the forecast of the 1989 P.M. Report and the annual bridge rehabilitation expenditures.

5.3.4 Capital Funds for Maintenance since 1996

Since 1996 FHWA recognizes certain maintenance tasks, such as full painting, as eligible for capital funding. The overall Federal fund allocation is not affected. The assumption is, therefore, that the increased maintenance expenditures will reduce the reconstruction needs. Although this claim has been repeatedly made, as in the P.M. report above, this effect cannot be achieved instantaneously. The reconstruction needs of New York City over the next 10 to 15 years will be entirely unaffected by an increase in maintenance activities beginning at the present.

The current point of contention between bridge managers and budget managers is to increase maintenance expenditures without reducing the reconstruction ones.

5.3.5 Bridge Maintenance Scope

The bridge management versus budget management argument formulated above puts pressure on bridge maintenance to demonstrate its effectiveness. While the deterioration rates at reduced or no maintenance are based on many years of experience, those at "full maintenance" are speculative. The curves of Figure 4 suggest that certain bridge components deteriorate and fail first and their malfunction accelerates the decline of decks and primary members which, in turn, determine the need to rehabilitate a bridge. Most notable among the rapidly deteriorating bridge components are joints, scuppers, wearing surface and paint. Their useful life appears to be between 5 and 10 years and their deterioration is known to accelerate that of bearings, pedestals, decks and primary members. The immediate task of full maintenance should be to conduct measures extending the life of these most sensitive components. Frequent spot-painting, washing, cleaning of scuppers and wearing surface repair are among the measures usually recommended. Designing new bridges to minimize their life-cycle cost by eliminating maintenance-intensive components, such as joints, is a relatively recent priority.

5.3.6 Bridge Maintenance Cost

Within the structure shown on Figure 3, adequate maintenance can be shown as the only effective means of improving the overall bridge conditions (Yanev 1996). Is that economically feasible?

A comparison of the "reduced maintenance" strategy and a proposed "full maintenance" strategy with their respective costs in 1997 U.S. dollars per square meter of bridge deck is illustrated on Figure 5, using current annual rates as shown in Table 3:

Table 3. Annual cost of bridge activities [$ / sq. m]
Full reconstruction................................4,300.
Component rehabilitation.........................1,616.
Full maintenance......................................20.
Reduced maintenance...............................12.
Reduced + Demand maintenance....................40.
User costs due to reconstruction.................$UC

It is assumed that bridges receiving minimum maintenance will deteriorate at the rate of 0.2 of a rating point/year. Below the rating of 4.00, there is need for "demand" maintenance, e.g. repairs to make the structure safe for the traveling public until reconstruction. The sum of the annual expenditures over 60 years without discounting is:

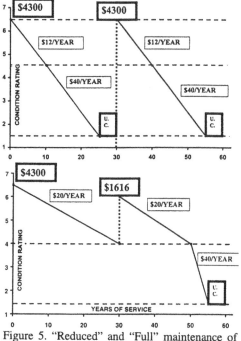

Figure 5. "Reduced" and "Full" maintenance of 1 sq. meter of bridge deck over 60 years of service.

$$2(\$4300+10\$12+15\$40+UC) = \$10,040 + 2UC \quad (1)$$

Alternatively, the new or rehabilitated structure can receive full maintenance at $ 20.00/sq. meter) and deteriorate at 0.1 of a rating point for 30 years, until it reaches the rating of 4.0. A component rehabilitation can restore the structure to a rating of 5.5 and it can, again, be fully maintained until it reaches the rating of 4.0 in approximately 18 years. The non-discounted sum of annual expenditures is:

$$\$4300+\$1613+48\$20+12.5\$40+UC=\$7,373+UC(2)$$

As discussed in (Yanev April 1994), the comparison of the two strategies represented by Equations (1) and (2) above is heavily dependent on discount rates and does not necessarily gain positive insight by introducing them. Table 4 compares the values obtained by Equations (1) and (2) at no discount with those obtained at discount rates of 4% (standard for the United States) and 8% (high).

Table 4. Effect of discount rates on maintenance strategy assessment in $/sq. meter of bridge deck

60 Years Service	Discount Rate		
Direct Costs	0	4%	8%
Reduced Maintenance	10,040	6,167	5,011
Full Maintenance	7,373	4,848	4,586
Reduced / Full	1.36	1.27	1.09
User Costs			
Reduced Maintenance	2UC	0.49UC	0.16UC
Full Maintenance	1UC	0.1UC	0.01UC
Reduced / Full	2.0	4.9	16

The comparison of Table 4 demonstrates that, as discount rates increase, the savings in direct costs due to maintenance expenditures become less significant. This has cast doubt on the *present worth method* as a tool suitable for such analysis. On the other hand, the same method indicates (correctly in this case) a significant increase in user costs due to reduced maintenance. User costs, however, are hard to estimate and rarely play a significant part in optimizing bridge management decisions.

It should be noted that:

- rehabilitated bridges are never fully restored to a condition rating of 7 and have been shown to deteriorate faster than new ones;
- the 0.1 deterioration rate at "full maintenance" is still an assumption based on the considerations stated in *5.3.5* above;

- user costs are incurred not only during bridge closures but also during reduced service due to poor conditions.

Experience has shown that rehabilitation work, such as the one shown after 30 years of reduced maintenance in Figure 5, is systematically missed by computer based analysis of bridge deterioration rates. As a result both strategies might be erroneously interpreted to suggest the same useful life of 60 years. This is suspected to have played a certain role in some past studies of this kind.

The term "full maintenance" has been used herein as coined by the 1989 P.M. Report. However significant developments have followed the completion of the latter, rendering the cost of "full maintenance", determined at the time, obsolete.

Most important among them is the rigidly enforced requirement to provide full containment during the removal of lead paint from bridges which are being re-painted. As a result the cost of bridge painting with full surface preparation has increased from $2.00 in 1989 to a current $15.00 per square foot. In view of this development and taking into account the expected performance of new paints, the 8 year full painting cycle recommended in the Report is now extended to a 12 year cycle. The lead paint removal is an expense unique over the life of a bridge but it is so significant that for small bridges a replacement may become preferable to repainting. Since 85% of the New York City spans are steel and all of their paint includes a lead based primer, the problem will be a persistent one.

Recognizing the above, the New York City Department of Transportation has commissioned a Columbia University based University Consortium to revise and update the Preventive Maintenance Report, starting in January of 1998. Among the tasks are the incorporation of the new paint protocol, as well as a recommendation for optimal maintenance of bridges which are neither new nor near rehabilitation. It should be noted that the Component Rehabilitation Program, not existent in 1989 and recommended by the original P.M. report, is fully functioning since 1993. Another point to be considered is the feasibility of eliminating rock salt as the only de-icer applied by the New York City Department of Sanitation to the City traffic network, including the bridges.

19

The updated P.M. Report is expected to act as a document substantiating the New York City maintenance needs and thus form a basis for the bridge management general policies of the near future. Nonetheless, funding decisions recommended by the City Bureau of Bridges will remain subject to approval by the Office of Budget and Management.

5.3.7 Unique Structures

Past studies have demonstrated (Yanev, 1994) that the unique structures should be treated independently from the rest of the City bridges for all purposes of forecasting needs and condition assessment. The above reference, for instance, showed that more than 50% of the emergency repair bridge work is conducted on the four East river bridges (Brooklyn, Williamsburg, Queensboro, Manhattan) and the 25 moveable ones.

The scope of optimal maintenance for the majority of City bridges is not entirely applicable to the unique structures. Also different are the requirements for the water crossings in general (10% of the total number).

The East River and most of the movable bridges are among the most heavily traveled ones and the most deteriorated. The East River crossings are currently under rehabilitation, the cost of which, over a period of roughly 20 years (past and future) is estimated as shown in Table 5:

Table 5. East River Bridges Rehabilitation Cost $m
Brooklyn $321.29
Manhattan.....................................$611.30
Queensboro...................................$447.70
Williamsburg.................................$697.21

The principal movable bridges are also under or awaiting rehabilitation. These major construction projects are funded jointly by the Federal, State and City Government. As a result, all participating agencies have the authority to supervise certain aspects of the scope of work. The Federal Government has made it a requirement that bridge specific maintenance manuals be part of the rehabilitation design. FHWA is also prepared to fund structural modifications enhancing the maintainability of a bridge. On the Brooklyn bridge, for instance, a sprinkler system is currently installed by the City with Federal funding in order to test its ability to spray anti-icing liquid over a stretch of the roadway surface.

The bridge specific maintenance manuals and their rigorous implementation will be a significant positive step in the funding and the management of the unique City bridges.

The most distinctly unique feature of suspension bridges, their high strength steel parallel wire cables, was recently addressed as follows. The four Metropolitan area owners of such bridges, namely the MTA Bridges and Tunnels, the Port Authority of New York & New Jersey, The New York State Bridge Authority and New York City Department of Transportation jointly let a contract under which researchers at Columbia University are to combine all previous research data for the suspension bridges of the four owners and to summarize and evaluate all findings. The resulting report is expected to advance the knowledge on suspension bridge cable performance and maintenance. The National Science Foundation and Research Council indicate that this study may serve as a first step in a nationwide survey.

Motivated by similar concerns and observing the United States experience, notably that of the Williamsburg bridge, the managers of the Akashi-Kaikyo bridge in Japan opted for a high first-cost maintenance intensive device, designed to introduce hot dry air under the suspension cable wrapping in order to minimize the risk of corrosion of the high strength wires.

Another feature, significantly affecting the management of unique structures is their possible landmark status. Such a status introduces important constraints to maintenance, rehabilitation and use as well as different funding mechanisms and life-cycle considerations which should be taken into account.

6 CONCLUSIONS

Bridge management will always remain a process of interaction between structures, vehicles, users, designers, builders, economists, politicians or, to various degrees, all members of the community. Consequently all conclusions are subject to future revisions. The proposed ISTEA 1997 was not approved by United States Congress before the winter recess. Further debate will be held in early 1998 in order to negotiate the distribution of

Federal funding for the transportation infrastructure among States and localities. The outcome will significantly influence bridge management nationwide. Despite the uncertainties built into the bridge management process, however, a number of conclusions, both general and specific, are currently emerging as noteworthy.

Bridge management and the computer-based BMS developed over the last decade are two distinctly different entities. In one form or another, bridge management is as old as bridges. The recent interest in BMS is part of the general effort to draw attention to certain deficiencies of bridge management, to clarify and improve the process of selecting optimal bridge management strategies under defined constraints.

In many computer based BMS the decision making module is either missing, under development or limited to allowing for the intervention of a qualified individual. This is likely to remain the case in the future. At best, BMS will be in a position to recommend certain types of action under certain constraints and to compute the penalties of failure to follow this recommendation.

The bridge management tasks are more readily met for smaller bridge networks. In recognition of this fact, FHWA has developed different BMS (PONTIS - for larger systems, BRIDGIT - for smaller ones). Nonetheless, the bridge network serving a large metropolis, such as New York City cannot be fragmented since all important decisions have to address the global consequences. Special considerations of unique structures or structural sub-networks have to be incorporated as details of the system.

The most significant contribution of the bridge management systems will remain their comprehensive and readily accessible database. The amount of maintenance and reconstruction work, currently missing in the databases would be a very valuable enhancement.

The sensitivity of bridge component condition to different maintenance levels is currently (partially or fully) unknown. Data on this important bridge management variable will soon begin to arrive and will have to be preserved with great attention. The types of maintenance most effective in extending the life of bridge components will vary depending on bridge types, local conditions, etc.

A recent development in the United States is the added constraint on steel painting. As a result the cleaning, washing of bridges, always recommended, gains much greater importance.

The quantification of bridge component conditions will remain a challenge for engineers with considerable area of improvement. Non-destructive techniques and on-line bridge monitoring are considerably improving the state of the art at very accelerated pace.

The overall average condition of a large bridge network is not a particularly sensitive, and therefore, significant indicator. Averaging data from a large number of bridges in order to determine the deterioration or condition patterns leads to false results, due to incorporating undocumented and unknown factors. Deterioration rates of bridges and bridge components, in particular, have been generated by a number of authors for various bridge networks. Most results have shown a convex curve, asymptotically approaching a certain rating higher than failure. There has been speculation about the reasons for such a pattern. Our latest observations (Yanev 1997) indicate that slowing down of deterioration patterns for bridges and bridge components with age is due primarily to undocumented repair work. The effect is compounded by subjective condition ratings which are quick to down-rate a new bridge but reluctant to close a poor one completely. A more realistic pattern is obtained by reviewing specific cases where the fastest known deterioration is observed and failure of certain components is documented.

The level of maintenance "optimal" for the bridges is always higher than the "optimum" within the more general budget structure of the metropolis. When the average age of bridges is past 30 years their average condition is bound to conform to a normal type distribution with the majority of structures clustered around the median of the range. The customary methods (such as the present worth) for "optimal" maintenance selection are not fully capable to produce a realistic long-term cost-benefit assessment. In addition to the complexities of such an analysis, the costs and benefits are incurred by different entities with different priorities. New developments likely to influence the funding optimization decisions of the future will include new types of construction. Jointless bridges with integral abutments, bridges with integral deck and primary members, pre-compressioning the deck, new painting systems and other innovations will significantly modify the cost relationship between construction and maintenance. The optimal life-span of bridges is likely to become increasingly dependent on structural type and will vary accordingly.

The current period in bridge management in the United States and, in particular, New York City is marked by strong efforts to direct major Federal and local funding to transportation infrastructure and, particularly, bridge maintenance. Given the ever present funding limitations, the source of maintenance funding is sought in reconstruction funding. The demand of the moment is to assure that bridge reconstruction needs are adequately met and that future bridge maintenance is conducted in a way which will effective reduce future reconstruction needs without adversely affecting the service of the infrastructure.

ACKNOWLEDGEMENT

This article presents views of the author and not those of any agency or organization. The assistance of the Bridge Management staff of the New York City Department of Transportation and, in particular, that of Ms. Paula Friend, Mr. Duong Tran and Mr. William Livingston was indispensable in preparation of the material.

REFERNCES

Federal Highway Administration 1988. Recording and Coding of the Structure Inventory and Appraisal of the Nation's Bridges. FHWA -ED-89-044.

Llanos, J. & Yanev, B. 1991. Models of deck deterioration and optimal repair strategies for the New York City bridges. In Proceedings, Second Civil Engineering Automation Conference: 1-28, New York November 19, 1991.

National Cooperative Highway Research Program: NCHRPR No. 300 1987. Bridge Management Systems. ISBN 0-309-045690-X.

New York City Department of Transportation, 1990, 1996. Bridges and Tunnels Condition Report

Organisation for Economic Cooperation and Development 1992. Bridge Management.

Robichon, Y. & Binet, C. & Godart, B. 1995. Evaluation of bridge condition for improved maintenance policy. In Extending the lifespan of bridges. IABSE symposium, San Francisco.

Transportation Research Records:
TRR No. 1183 Systematic Approach to Maintenance 1988. ISBN 0-309-04723-4.

TRR No. 1184 Structures Maintenance 1988. ISBN 0-309-04734-2.

TRR No. 1290 Third Bridge Engineering Conference 1991. ISBN 0-309-05067-7.

TRR No. 1490 Management and Maintenance of Structures 1995. ISBN 0-309-06154-7.

Yanev, B.S. & Chen, X. 1993. Life-cycle performance of New York City bridges. In Transportation Research Record No. 1389, Materials and Construction: 17-24, ISBN 0-309-05459-1.

Yanev, B. S. 1994. Emergency repair needs assessment for the New York City bridges. In Maintenance of bridges and civil structures, Colloque International: 501-516, Paris, 18-20 October 1994, Presses de l'Ecole National des Ponts et Chaussees.

Yanev, B. S. 1994. User costs in a bridge management system. In Characteristics of Bridge Management Systems: 130 - 138, Transportation Research Circular #423, April 1994, ISSN 0097-8515.

Yanev, B. S. 1996. Optimal maintenance and rehabilitation strategy for the bridges of New York City. In Proceedings, International Conference on Retrofitting of Structures: 311-327, Columbia University, New York.

Yanev, B. S. 1997. Life-cycle performance of bridge components in New York City. In Proceedings, Recent Advances in Bridge Engineering: 385-392, US - Canada - Europe Workshop, Zurich, July 1997.

Operation and Maintenance of Large Infrastructure Projects, Vincentsen & Jensen (eds)
© 1998 Taylor & Francis, ISBN 90 5410 963 7

Transport Telematics: An indispensable tool for an efficient operation and maintenance of the transport infrastructure – Contribution from the European Union

W. Maes
European Commission, DG XIII (Telecommunications, Information Market and Exploitation of Research), Brussels, Belgium

ABSTRACT: Transport Telematics as part of an intelligent transport system, can help optimise existing transport infrastructure capacities, link networks, improve safety and reduce the negative effects of road transport on the environment. It can also provide logistical support for transport operations and improve the efficiency of collective transport operations. The growing Information Society will also benefit transport users by providing a range of "added value" services. And at the same time, market opportunities for industry and service providers will be created.

Since 1988 the European Commission launched several R&D programmes to develop transport telematics in Europe. In 1997 it also issued a Communication on a general strategy and framework for the deployment of telematics in the road transport sector.

An overview is given of the main achievements in the R&D programmes as well as the actions proposed to accelerate the deployment of road transport telematics in Europe

1. INTRODUCTION TO THE INVOLVEMENT OF THE EUROPEAN COMMISSION IN THE FIELD OF TRANSPORT TELEMATICS

The success of European Union integration has created an enormous expansion of trade and mobility of people and goods. But it has also resulted in an unbalanced and overburdened transport system, with congested roads, high pollution and accident levels, and losses of efficiency.

The previous policy of trying to keep up with transport demand by concentrating on increasing capacity and expanding infrastructure has rightly been abandoned as unsustainable. In its place, Common Transport Policy now stresses the need for intermodal systems and economic instruments to re-balance transport for 'sustainable mobility'.

The development of transport telematics is a crucial mechanism for carrying out this new dimension of transport policy. The successful application of transport telematics makes use of technology to improve the movement of people and goods and offers significant opportunities for increased transport efficiency, better safety and comfort for passengers, and less pollution in the environment. On the other hand the European Union is pushing ahead with liberalisation in the telecoms sector which will lead in the short term to an explosion of new services for the citizens. The convergence o computing and telecommunications as well as with the broadcasting and publishing sectors leads to a rapid evolution of the Information Society and the development of citizen services such as new types of transport telematics services.

Specifically, telematics now more than ever before will provide traffic management, information services and intelligent on-board systems which can:
– improve the management of existing transport networks
– facilitate the integration of different transport modes and services, and support modal shifts
– improve traffic safety
– provide better traffic flows and thus reduce environmental pollution

Telematics will also
– facilitate the provision of high-quality added-value transport services for public transport and professional operators
– provide tele-services moderating the demand for transport.

The European Commission is committed to the successful implementation of a transport telematics solution. Since 1984 EU research and technological development activities have been implemented under a series of multiannual Framework Programmes containing several specific programmes. The first EU specific programme in the domain of transport telematics was named DRIVE (1989-1991). It was closely linked to the EUREKA project PROMETHEUS developing the intelligent car and was followed by DRIVE II or the Advanced Transport Telematics part of the "Telematics Services of General Interest" Programme (1992-1994). In the fourth Framework Programme the sector of Transport Telematics makes part of the Telematics Applications Programme (1996-1998). While DRIVE mainly focused on road transport and DRIVE II on road and intermodal transport, in the fourth Framework Programme telematics applications in all transport modes as well as intermodal transport are under consideration. The European Commission's contribution to these different transport telematics programmes amounts to more than 400 MECU. In the mean time a proposal has been made by the European Commission for the 5th Framework Programme (1998-2002). In this proposal advanced intelligent systems for transport and tourism are recognised as priority items to be considered under the key action "Systems and services for the citizen" aiming at "Creating a user-friendly Information Society". Strong links exist with another key action in the same proposal on "Sustainable mobility and intermodality" aiming at "Promoting competitive and sustainable growth" .

So far, the achievements of 70 projects in the Transport sector of the Telematics Applications Programme globally indicate that the *telematics technologies* for:

- location systems (Global Positioning System, GPS; Global Navigation Satellite System, GNSS)
- communication systems (Global System for Mobile Communications, GSM; Dedicated Short Range Communications, DSRC)
- information systems (Radio Data System - Traffic Message Channel, RDS-TMC; Variable Message Signs, VMS)
- in-vehicle guidance systems (autonomous, infrastructure supported)
- personal intelligent communicators (Personal Traveller Assistant, PTA)
- control systems (Urban Traffic Control, UTC, bus priority, public transport operation)
- electronic tolling (smart cards), etc.

and the *tools* to operate these such as:

- digital maps
- data dictionaries
- protocols (ALERT-C; Traffic Message Channel, TMC)
- data exchange (EDIFACT, DATEX)
- message exchange formats between Traffic Information / Control Centres (TIC, TCC)
- Public Transport data models
- etc.

are ready for implementation. These technologies and tools provide the basis for interoperability and transferability of telematics applications for transport between and within the Member States. They contribute to sustainable mobility and to the solution of transport problems in the application areas of traffic information and control as well as transport management on the Trans-European Networks (TEN) and in the metropolitan and rural areas addressing all modes, i.e. road, rail, air, and waterborne transport.

Apart from the research and development activities, also general strategies and frameworks for deployment have been prepared, starting with the road transport sector, with such applications as the RDS-TMC based traffic information services and electronic fee collection (see chapter 5.)

2. TRANSPORT TELEMATICS FOR CITIZENS

2.1 *Transport in the Information Society*

The convergence of information and communications technologies, as well as the audio-visual media and the publishing sector, is sweeping away the traditional boundaries of electronic applications and replacing them with information highways which are changing the whole way we live, work, and trade. The vision of an Information Society - with more efficient production and distribution of goods and services, higher levels of education and skilled-job opportunities, and a better quality of life for everyone - is fast becoming a reality. And the role of transport in this revolution is vital.

Passenger transport is a major challenge for Europe. In the world's most concentrated urban region, demand for passenger transport is high, and growing rapidly. Private cars have satisfied this demand up to now, providing 80% of journey needs on an ever-expanding network of roads, but at a price of unacceptable levels of traffic congestion, pollution, accidents, and loss of efficiency.

2.2 *A new integrated approach is needed*

With the number of cars in the E.U. perhaps doubling from the present 156 million by the year 2010, with an ageing population having more free time, as citizens demand more and more freedom and easier access to work and leisure, urgent measures are called for.

The European Union is responding to this challenge with the development of an integrated and intermodal approach to public transport needs (the "Citizen's Network").

It is designed to enable people to travel on a variety of transport systems and to switch easily between them.

2.3 *Based on new telematics applications*

Key to this policy is the deployment of a new generation of telematics applications which will facilitate the availability, efficiency, and safety of an integrated transport system and the delivery of information to the users.

For this reason, the European Union, at both national and community levels, has undertaken a significant programme of research/development and supports the implementation to ensure reliable and efficient transport management and information systems which can provide "Mobility for Everybody".

Projects of particular importance to the travelling public, include:

2.3.1 *Information for all, everywhere, at any time*

Universal travel information covering all types of transport should be accessible to Europeans when and where it is needed, if a future integrated transport system is to succeed. A series of development projects focus on the construction, content, and delivery of public and private transport information to all sectors of the population, young and old, urban and rural, including disadvantaged and disabled groups. This information is delivered via fixed terminals but also through hand-held portable devices using mobile communications.

The greying of Europe poses special questions for transport. 50% of European adults will be aged 50 or more by the year 2010. Many will be mobile and affluent, others will not. Changing transport habits and adapting to new information technology will not be easy. They will need comfortable, reliable, safe transport and information which is easy to access and simple to understand - delivered when they need it, in the home, at the workplace, and on the move.

2.3.2 *Information to drivers and payment systems*

The private car usage will undoubtedly remain a dominant feature in future integrated transport systems. But the deployment of telematics to deliver in-car information can greatly increase efficiency and safety and reduce traffic congestion and pollution.

Projects aiming at validating RDS/TMC (Radio Data System/Traffic Message Channel), a system delivering radio traffic information in the driver's chosen language, are well advanced. National and Euroregional projects are being pursued in 11 European countries. A Memorandum of Understanding has been signed by most of the Member States as well as some private organisations to use common functional specifications in order to guarantee interoperability and continuity of this service cross-border.

Other projects led to the development of travel and traffic information services including emergency calls on the pan-European mobile telephone network GSM.

To survive on the market, all of these systems must be commercially viable and tests have involved car drivers, service providers, broadcast companies, and road operators to assess costs, benefits and market potential. Results have been positive and drivers have shown high acceptance.

Other telematics applications for general usage in road transport include electronic tolling and ticketing payment systems. By using smart cards for automatic payment, public transport travellers and drivers paying tolls find their journeys easier. Widespread adoption is expected in the near future. A Council Resolution has been adopted in 1997 on the development of telematics in road transport, in particular with respect to electronic fee collection (EFC). In this resolution the Council calls on Member States and the Commission to develop a strategy for the convergence of EFC systems in order to achieve an appropriate level of interoperability at a European level, taking into account systems already existing and the work in the European standardisation bodies.

2.3.3 *Driver assistance*

Road safety is a key objective for all passenger vehicle projects and specific projects have been set up to develop telematics applications which can improve driver and passenger safety. Improving driver comfort and environmental performance are additional benefits.

Main projects focus on GSM based services to get help in emergency situations; collision-avoidance systems, such as radar applications increasing the gap between speeding vehicles; and advanced autonomous intelligent cruise control, which enables vehicles to travel closer together with controlled safety distances, thus reducing congestion and increasing safety. Short-wave communications technology is being developed to deliver speed warnings to the driver from the infrastructure side.

3. TRANSPORT TELEMATICS FOR CITIES

3.1 *Planning for sustainable mobility*

Urban areas, where 80% of Europeans live, dominate the economic and social development in the Member States of the European Union. But the increasing use of private transport is now paralysing these cities and threatening the living standards and quality of life of their inhabitants.

Over the last 25 years passenger transport volume has increased by 85%, most of it away from collective transport to the private car. In cities, the collective transport share is generally improving but is growing less than the private car does.

This growth in car usage has had two major effects. Firstly, it has encouraged the increase of distance between the home and work, shopping and leisure locations, further increasing the demand for private individual travel. Whereas people used to live near their work, and typically use one-stop public transport, separation has now become accepted. Secondly, it has led to massive urban congestion, daily pile-ups in city entrances, an unacceptable level of accidents and pollution, and loss of efficiency.

3.2 *From car domination to integration*

This is the background against which the European Union has developed its "sustainable mobility" policy, based on integrated transport systems using a mix of transport modes. It is not the policy to banish the car - though this may happen in some city-centre areas - but to integrate the different modes of transport in order to provide real alternatives to the car. From car domination to system integration, for a more rational use of private vehicles.

For an integrated approach to work successfully, local and national authorities must put passenger needs at the centre of their policy. And public transport must be more accessible, safer, cleaner, and more affordable, if it is to attract increased usage. The long-term goals must be to lift the quality of public passenger transport management and set up an efficient passenger system that enables travellers to switch easily between modes.

To meet this objective, the European Commission has set up programmes of R&D and implementation support projects; most of the projects are based on a new generation of advanced applications to provide information services and transport management.

3.3 *Urban traffic management*

Foundations were laid back in 1988, when a number of European cities were grouped together under the POLIS network to study the problems of urban transport and to provide transport telematics solutions which could be applied Europe-wide. Today, "smart" urban systems - based on intermodal travel and collective transport - are being developed for the next century.

Transport telematics is playing a key role in helping urban authorities strike an efficient balance between travellers' needs and the capacity of transport networks. Some European cities are deploying integrated telematics systems for congestion management, traffic and parking control, and reduction in car usage.

New traffic monitoring technologies are measuring total traffic flows, flow composition, queuing, accidents and speeds. Control systems, using real-time traffic information, are being used to adjust traffic signal timing, to set priorities and to speed up flow especially for public transport. Variable message signs (VMS) have been evaluated to better understand the driver response to the information provided.

Access control, using automatic vehicle identification, has been successfully tested in city centres; with big reductions in travel time and pollution output, and a corresponding increase in public transport usage. Other information systems are being applied to improve freight flows in inner-urban areas.

3.4 *Public transport*

Projects have focused on the identification, development, and promotion of technical and organisational solutions which can improve the appeal and efficiency of public transport:

* passenger information services, including real-time on-board information as well as at bus stops, trip-planning terminals, and demand-responsive services in low-demand regions
* fare collection and integrated ticketing prototypes, including smart card technology applied to electronic fare management
* automatic vehicle monitoring, scheduling and control systems (AVM/VSCS) and bus priority measures for operators
* satellite systems for navigation, communications and control.

4. TRANSPORT TELEMATICS FOR EUROPE

4.1 *The need for a Europe-wide approach*

The interdependence of national economies in the European Union and increasing cross-border travel by its citizens for work and leisure, has dictated the need for an international and integrated transport policy. Local systems which falter at national frontiers make no sense. Europe-wide solutions are the rational way forward.

The European Union has therefore invested heavily in trans-European infrastructures to ensure that the movement of goods and passengers within its boundaries - and beyond - is more reliable, faster, safer and environmentally sustainable.

4.2 *The need for advanced telematics solutions*

The development of transport telematics is central to the Union's Transport, Telecoms, Industrial and Information Society policies. Through advanced and high-quality services, people and businesses will enjoy more efficient and convenient highways, airports, railways, waterways and ports; and linkages between different modes of transport. European industry will benefit from new products and citizens from reliable information services.

The European Commission has led from the front and managed a vast research programme across the entire telematics chain; from data capture and processing to transmission and reception, right through to standardisation and common practices. With emphasis on user needs, inter-operability, and ease of access, the private and public sectors have combined forces to provide transport systems and services which will take us into the next millennium.

4.3 *Long-distance road transport*

The strategy for improving long-distance road transport is based on traffic management and traffic information services, as well as freight and fleet management services using transport telematics tools currently in development, testing and demonstration.

Traffic management focuses on monitoring and controlling traffic flows, and responding to emergencies due to weather conditions or accidents. The objective is to reduce congestion, pollution and economic wastage by improving flow and safety. A range of data-exchange systems are being developed to ensure the continuity of traffic information and management services on the 70 000 km Trans-European Road Network (TERN) and other busy inter-urban connections.

Traffic information services tell drivers what's happening on the roads so that they can plan their journeys better and make routing choices. Work undertaken by the DATEX task force has established an interoperable traffic data exchange protocol for messaging across the existing European systems. A Memorandum of Understanding has been signed by most of the EU Member States and some private organisations to guarantee the adherence to this protocol by all actors involved at least on the TERN.

Freight and fleet management services help to ensure the most efficient use of the available freight transport system by improving freight transport reliability, safety and customer service and ensuring better conditions for drivers and dispatchers. They also help reducing the unfavourable effects of freight transport on the environment and ease congestion on the roads.

4.4 *Intermodal goods transport*

Road transport's increasing share of European freight and passenger transport is choking the highway system and causing unacceptable levels of pollution. Current growth projections for trade and private car usage show the situation will only get more difficult.

European policy now recognises intermodal transport as a solution. It aims for a more balanced use of existing transport capacity by upgrading environmental friendly modes such as railways and increasing the coherence between alternative services, right across Europe. But efficient intermodal transport needs accurate real-time data, making the use of telematics technology essential.

Projects focus on tracking and tracing technology to

identify and manage cargo in multi-modal situations. A specific goal is the integration of information systems specific to each transport mode, especially for container-handling throughout the transport chain.

Development work addresses the integration of modes at three levels:
- infrastructure and transport hardware
- operations and terminal use
- services and regulations.

4.5 Rail

The European Commission believes that railways should play a much greater role in tackling Europe's transport problems and is committed to achieving this goal. With investment and initiatives which respond to customer needs, rail transport could do much to provide sustainable mobility for goods and passengers in the future, both nationally and throughout Europe.

Research projects have concentrated on overcoming some of the major existing weaknesses, such as the absence of a network for data interchange along the intermodal transport chain. The absence of European standards for rail-traffic management systems is another serious handicap which is being addressed; first on high-speed networks, later on conventional rail.

A European strategy to revitalise railway systems has been developed and freight freeways using existing rail infrastructure are designed. In addition, the Community research programme is laying the foundations for co-operation between Member states to revitalise rail in Europe, with a series of projects creating new traffic management and information services (including intermodal systems), and the ERTMS project which is providing harmonised command/ control across frontiers.

4.6 Satellite navigation

As non-military use of US and Russian satellite-based navigation systems grows, and the cost of receivers falls, applications for improving European transport performance are taking shape. In-vehicle receivers, combined with other communications systems, will soon be standard equipment. Precise driver guidance information, displayed on digital maps, will produce freight-movement efficiencies and speed up passenger traffic, goods delivery and emergency services.

Europe is now actively involved in the development of the next generation of global navigation satellite systems (GNSS) to provide highly-accurate navigation and position-fixing services for road, rail, maritime and aviation applications. A Commission Communication on this matter has been issued on 21 January 1998.

5. DEPLOYMENT OF ROAD TRANSPORT TELEMATICS IN EUROPE

5.1 Action Plan

Road Transport Telematics (RTT) gives policy makers a complementary tool to road building.

By providing safer driving with fewer delays for individual travellers, logistical and management support for transport service providers and more effective traffic control for road operators, RTT can help make European roads safer and more efficient. It can also provide new business opportunities for providing new types of user oriented services and the development of new products.

This has been recognised by all decision-makers, right up to the Council of the European Union and the European Parliament. The Commission, assisted by the Road Transport Telematics High Level Group constituted of representatives from the Member States, has therefore prepared a general strategy and framework for the implementation of telematics in the road transport sector and a rolling action plan for deployment up to 1999. It is clear that this deployment will only happen if there is a strong private sector participation and funding, including public/private partnerships.

The action plan sets the following priority areas:

Traffic Information Services (RDS-TMC) to provide a broadcast service to drivers with messages in the language of their choice on traffic conditions via in-vehicle audio or visual displays. The R&D phase is almost finished and standardisation for cross-border interoperability is progressing well.

Electronic Fee Collection (EFC) covers toll collection and road price differentiation services. The technology is proven; EFC pre-standards are adopted and a convergence strategy for achieving interoperability in Europe will be developed taking into account the systems already existing and the work in the European standardisation bodies .

Transport Data Exchange and Information Management systems are needed for almost all RTT services. Action is concentrating on technical standards and operating protocols as well as on a framework to enable a take off of commercial value added services.

Human Machine Interface (HMI) concerns telematics devices inside vehicles, for both information displays to support driver's decisions and vehicle control systems. Continuous R&D work is being conducted and a taskforce has been set up to establish a Code of Practice on HMI of in-vehicle information devices.

System Architecture: the aim here is to define a European open system architecture in order to achieve an appropriate level of interoperability for all RTT systems and applications introduced .

These five applications are a first step in the implementation process of RTT in Europe. Other priority applications, which need still further analysis to define if and what action is required at European level, are pre-trip and on-trip information and guidance, inter-urban and urban traffic management, operation and control, collective transport, advanced vehicle safety/control systems and commercial vehicle operations.

In accordance with the principle of subsidiarity, the EU strategy will build on actions at Community level where progress can best be achieved by Community action, or where success cannot be sufficiently achieved by action at national level. Member States will develop their own deployment strategies for RTT, reflecting national priorities, requirements and preferred legal and institutional arrangements. In particular they will need to determine, which RTT applications should be provided as a service of general interest (however delivered, whether by the public sector, the private sector or both in partnership) and what is for purely commercial exploitation by the private sector.

5.2 *EC deployment support*

In order to promote the deployment of transport telematics the European Commission also supports projects of common interest via the budget lines for the Trans-European Networks for Transport and Telecom.

5.2.1 *Trans-European Transport Networks*

For pan-European transport, the Community's Common Transport Policy is based on the development of the Trans-European Network (TEN-T) concept to provide an integrated network which can strengthen economic and social cohesion and provide safe and environmentally-sustainable mobility.

This TEN-T network will bring land, sea and air transport infrastructure networks together and link Europe from east to west and north to south. The policy guidelines also require that TEN-T has the capacity to be connected to the networks of Central and Eastern European countries, with a set of multi-modal corridors and to Mediterranean rim countries.

High-speed road and rail trans-European networks are a prime focus for transport telematics infrastructure and services, and TEN-T policy foresees the deployment of road traffic management and rail command/control systems as key to the whole project.

5.2.2 *Trans-European Telecommunications Networks*

The establishment and development of trans-European telecommunications networks aim at ensuring the circulation and exchange of information across the Community. This outlay of equipment is a precondition to enable citizens and industry, especially Small and Medium Enterprises in the Community, to derive full benefit from the potential of telecommunications so as to make possible the establishment of the "Information Society", in which the development of applications, services and telecommunications networks will be of crucial importance with a view to ensuring the availability for each citizen, company or public authority, including in the less developed or peripheral regions, of any type or quantity of information they may need.

Full advantage will be taken of trans-European telecommunications networks to provide user-oriented services, in the domains of logistical support for transport industries and of the development of value added services, such as information services, integrated payment and reservation services, trip planning and route guidance, freight and fleet management and services in urban areas.

5.3 *Institutional Consequences*

Both the interoperability and transferability of European systems and services lead to the clear requirement to rethink and to finally redefine the institutional framework for the operation, not only in the individual countries but in a real European co-operation. Interoperability and transferability of services provide a wide range of opportunities to enlarge the possibilities for e.g. transport managers and operators and finally the end users in the field of traffic management and information when barriers are recognised and the institutional framework conditions are adapted in time.

For the development and deployment of facilities such as information services in many Telematics Applications Programme projects, public/private-partnerships have been set up in order to:

- make the best use of existing data sources and "content" which are still mainly owned by public administrations and
- organise the information provision and service operation and provision to the end users which is mainly the domain of private companies.

In some European countries or regions the authorities are in the process of defining and agreeing the institutional framework conditions for such a partnership co-operation, e.g.:

- In the region of Paris (Ile-de France) public traffic information servers have been linked - also with the server of the city of Paris - in order to finally provide information services to users via a dynamic Variable Message Sign system, the Internet and a commercial service called VISIONAUTE. A public/private partnership agreement has been made.
- In Germany, the Federal Ministry of Transport has initiated a telematics working group which now encourages the states ("Länder") to develop and agree the institutional framework conditions on e.g. the data exchange business between the public side (providing "conventional" traffic data from control centres) and the private side (offering "floating-car data" from their dynamic information services under deployment).

More examples could be surely given from other countries which would all indicate that there is a clear need for a European co-operation in defining a common institutional framework for the deployment and operation of systems and services in particular in the field of traffic information - the more transferable and interoperable services will become available.

6. BENEFITS OF TRANSPORT TELEMATICS

6.1 *Encouraging results and real benefits*

European Commission led research and development programs, in particular, DRIVE and the follow-up ATT (Advanced Transport Telematics) Programme, have produced a solid foundation of Research and Development and identified the benefits which are available from more intelligent use of our transport infrastructure.

These benefits include increased safety, reduced accidents, less congestion, time savings, and a transport system which is more compatible with the environment.

Based on an assessment of results from the 1992-1994 ATT projects, the following benefits or impacts illustrate the contribution to safety, efficiency and the environment.

- Safety;

  - Variable Message Signs integrated as part of a weather traffic management system which provide up-to-date accurate information to drivers on conditions have: reduced accidents by 30%; reduced the number of people killed or injured by 40%; reduced speeds by 10%; reduced accidents during rain by 30%; and reduced accidents on foggy days by 85%.

  - Emergency call systems using satellite and cellular radio technology which enable help to be summoned quickly in the event of an accident have: reduced the response time of emergency vehicles by 43% giving an increase in survival rates of 7-12%.

  - Driver Monitoring systems installed in commercial vehicles resulted in a 41% reduction in accidents and reduced accident severity.

  - Safer driving and reduced risk of accidents with Collision Avoidance Systems and Intelligent Cruise Control, which help drivers react to hazardous situations.

- Efficiency;

  - Variable Message Signs, which provide up-to-date accurate information to drivers, have shown up to 20% reduction in traffic delays.

  - RDS-TMC (Radio Data System - Traffic Message Channel), the system which provides drivers with easy to understand information on current traffic conditions via their radios is estimated to provide savings in travel times of 3-9%.

  - In-vehicle Route Guidance systems, which provide drivers with turn-by-turn instructions based upon prevailing traffic conditions to guide them to their destinations, are estimated to provide savings in travel times of 4-8%.

  - Control strategies based on information from Ramp Metering, which monitors the traffic flow joining and leaving major arterial roads, has led to increases of mean speed of 21% on motorways, 16% on parallel arterial roads and 19% over the corridor.

  - Fully integrated Urban Traffic Management Strategies which combine Urban Traffic Control (which automatically adjusts the phasing of traffic signals to optimise traffic flow), Traveller Information (which provides information on public transport options), and Public Transport Priority (which minimises delays for public transport by adjusting traffic signal phasing), produce an estimated 25% reduction in travel time for all modes of transport.

- Providing travellers with up-to-the-minute travel information via public access terminals influenced 60% of travellers who considered the effect that this information might have on their journey.
- Automatic tolling systems have shown time savings of 40 hours per year for the average motorway commuter.
- Freight and Fleet management systems based on telematics services provide an estimated 5% saving in travel time, 12% saving in despatch time and a 6% reduction in distance travelled.
- Environment;
  - Estimated reduction of CO emissions by 10%, HC by 5% and NOx by 5%, resulting from an observed 20% reduction in traffic delays via effective Variable Message Sign guidance strategies.
  - In-vehicle route guidance systems providing positive turn-by-turn instructions for drivers show a 5% reduction in fuel consumption.
  - Urban traffic control systems providing priority at traffic signals for public transport vehicles result in a 4-6% reduction in fuel consumption and emissions.
  - Freight and fleet management systems show a 4.4 % annual reduction in fuel consumption.
  - Telematics based access control systems show reductions of up to 50% in pollution within controlled zones, with a corresponding increase of only 13% in the surrounding uncontrolled zones.

These benefits fully justify the growing confidence in transport telematics and the investment made by the European Commission and those involved in the programmes. It is due to the positive achievements coming from the projects that increasing industrial, commercial and political attention is being paid to transport telematics in Europe.

6.2 *Social and Individual Acceptance*

Public support, acceptance and satisfaction with transport telematics is extremely high. Surveys conducted during the ATT Programme show that when given the opportunity to use transport telematics, user feedback is overwhelmingly positive.
Examples of the results obtained include;

6.2.1 *RDS-TMC (Radio Data System – Traffic Message Channel)*

Up to 300 messages can be provided every 15 minutes, users have immediate access to traffic information (they do not have to wait for traffic bulletins).
- 70% of surveyed drivers were satisfied or very satisfied by the service.

- In one trial 34% of drivers changed their route after receiving congestion-warning information, while 33% reduced speed (earlier) when approaching an accident of which they had been informed.
- 92% of the drivers supported RDS-TMC implementation in Europe.

6.2.2 *In-vehicle route guidance*

In-vehicle route guidance systems provide drivers with turn-by-turn information to navigate them to their destination. These systems have the potential to be linked with up-to-date traffic information to provide dynamic route guidance, which takes the prevailing traffic conditions into consideration.
- Surveyed drivers indicated that dynamic route guidance is capable of changing their driving habits either by leading them to take different roads or by enabling them to drive at different times.
- 55% of the test drivers considered the in-vehicle route guidance useful for travel time reduction.
- 38% of test drivers believed the recommended route was better than their own choice.

6.2.3 *Road side information*

Variable Message Signs, which provide drivers with traffic and travel information, have provided immediate benefits for drivers.
In Scotland 82% of drivers regularly using the (test area) routes indicated that they will follow the VMS information even if it is in conflict with other sources. In Amsterdam 80% of interviewees found the information to be correct, 98% understood what the sign meant, 68% thought that there was some improvement in driver comfort and 63% reacted to the information.

6.2.4 *Public Transport Information*

Real-time information at bus stops providing details of next bus, waiting time etc.
- 81% found the information to be useful
- 58% would support further investment
- Passenger anxiety reduced
- Passenger perception of service quality and the service provider improved.

While the examples provided illustrate the positive reaction of end-users when interviewed or completing questionnaires, they only relate to individual systems, which the participants have experienced or been made aware of. The problem of determining individual acceptance of transport telematics is that, for the most part, people do not realise they are using it.

Individuals do not knowingly purchase or use transport telematics, this is why the average citizen will not recognise this term. What an individual actually buys is a new car, motorcycle, or bicycle, a mobile telephone, a trip or journey, a ticket, some goods etc. While they may notice improved information or an attractive feature they tend to become aware of this as part of the package involved rather than as a specific item. An example of this is the success of TrafficMaster in the UK which has been greatly increased by the information delivery system (for traffic congestion information service that they offer) being installed as standard equipment on ranges of new cars.

Projects in the current EU R&D programme are already addressing this issue, for instance the HANNIBAL project raised the profile of transport telematics considerably during the 1997 World Alpine Skiing Championships in Sestriere, Italy, where they demonstrated a mobility management system to co-ordinate travel to the events. With co-operation and support from the Commission and from other projects the Championship was used to successfully demonstrate the advantages of transport telematics to the visitors, the authorities and to some 5 million television viewers.

7. CONCLUSIONS

Transport telematics has become an indispensable tool for an efficient operation and maintenance of the transport infrastructure. Public and private sectors have both an important role to play in the deployment of the new tools becoming available. Partnerships between both sectors are needed to take fully benefit of these tools. The European Commission via the research and development programmes and support given to the trans-European Transport and Telecoms Networks is making an important contribution to the accelerated launching of new services, and as such to the solution of the ever increasing transport and employment problems.

ACKNOWLEDGEMENTS

The author wishes to thank all his colleagues in the European Commission for their help in preparing this paper as well as the partners in the horizontal support action projects for the use which has been made of their reports as input to this paper. The contents of this paper reflect the view of the author and do not necessarily reflect the official views or policies of the Commission of the European Communities.

Operation and Maintenance of Large Infrastructure Projects, Vincentsen & Jensen (eds)
© *1998 Taylor & Francis, ISBN 90 5410 963 7*

Development of a bridge management system in Germany

Joachim Naumann
Federal Ministry of Transport, Bonn, Germany

ABSTRACT: Along Germany's federal trunk roads, which have a total length of around 55,000 km and constitute around 25 % of the total road network, there are currently 34,630 bridges. The total area of these bridges comes to nearly 24 million square metres. Although around 80 % of these bridges were built after the Second World War, the maintenance of the structures will be one of the major tasks of the future. In order to ensure the structural stability, traffic safety and proper functioning of bridges in the long term too, a Bridge Management System (BMS) is currently being developed, which is designed to implement purposeful maintenance strategies by intensifying the use of electronic data processing. The BMS is composed of seven modules, which are subdivided into several subject groups. This paper will explain the modules and describe the status of their development.

1 INTRODUCTION

Because of the totally different political and economic developments in West and East Germany after the Second World War, the stock and condition of the road networks in the two halves of Germany had also developed in totally different ways until reunification in 1990. In West Germany, work began relatively quickly on reconstructing the road network, and the economic boom meant that it was soon possible to substantially enlarge this network. In East Germany, on the other hand, there was largely stagnation; although the existing road network was gradually restored - admittedly much later than in the West -, there was hardly any new construction work or network enlargement.

Thus, when Germany was reunited, the road construction tasks and priorities were totally different in the two halves of Germany.

In the West, with the federal trunk road network with a total length of around 55,000 km, the objective identified in the requirement plans had been largely achieved. The tasks to be performed were shifting to the construction of roads supplementing or enlarging the network, the construction of bypasses and the elimination of accident black spots. At the same time, it was becoming apparent that the maintenance of the existing network would become increasingly important, despite the fact that most of the roads were relatively new.

In the East, on the other hand, the most urgent priority was, in the short term, to take immediate action to make the existing network able to cope with the sharp increase in the volume of traffic, and in the medium term to satisfy the enormous requirement for the construction of new roads and the upgrading of existing ones after decades of neglect.

In addition to the improvement of the main arteries, there was also a lot of work to be funded and realized on the secondary road network, the maintenance of which was severely neglected during the pre-unification era.

Given these great challenges, and the current budgetary restrictions caused by the economic downturn of recent years, a high degree of priority has to be given to the use of intelligent management systems for the optimum distribution of resources, especially to preserve the stock of existing roads. A major component of this is the development of a bridge management system.

2 BRIDGE STOCK

In the network of federal trunk roads (motorways and federal highways), which constitutes around 25 % of the total network, there are a total of 34,632 bridge structures (as at 31 December 1996) (Fig. 1). The stock also comprises 120 tunnel structures with

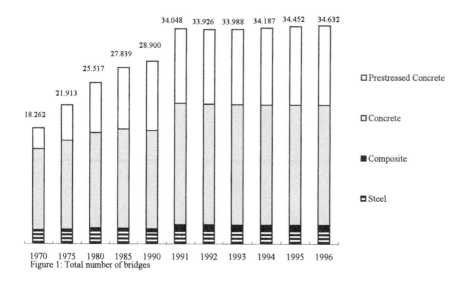

Figure 1: Total number of bridges

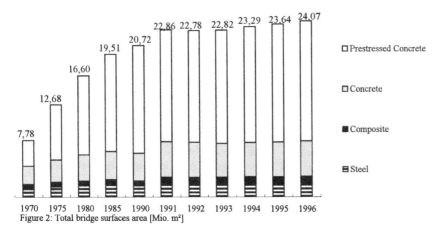

Figure 2: Total bridge surfaces area [Mio. m²]

a total tube length of 79.8 km, as well as a large number of retaining walls and noise abatement walls.

As a result of the reunification of Germany, the number of bridges on federal trunk roads rose by around 5,150, from 28,900 to 34,048. The fact that there has been a relatively small increase in the number of bridges since then is due to two factors. First, there has been less construction activity in the old federal states, and in recent years some federal trunk roads have been downgraded to secondary roads. Second, the construction of new roads in the new federal states has not yet made itself felt to its full extent. Here, a significant increase in the number of bridges is only apparent as of 1996 as a result of the completion of new sections.

Of the 34,632 bridges, 18,994 (55 %) are reinforced concrete bridges, 12,642 (37 %) are prestressed concrete bridges, 2,163 (6 %) are steel bridges and 833 (2 %) are steel composite bridges. There are very few masonry bridges in the federal trunk road network. The total area of the bridge stock is around 24.06 million square metres, with the largest share - 16.43 million square metres (68 %) - being constituted by prestressed concrete bridges (Fig. 2).

Because of the topography in Germany, the percentage of large bridges is relatively low. Only 2,057 bridges on federal trunk roads have total spans between their end supports with a length of over 100 m, which is a mere 6 % of the total stock. Over the years, a large number of different bridge types have been designed; standardized bridge types have only been used on a very small scale.

Because of the developments mentioned above,

there are great differences between the age structures of the bridges in the two halves of Germany. At the time of reunification, around 76 % were less than 30 years old in the West, whereas in the East almost 60 % dated back to the pre-Second World War era. As far as bridge management is concerned, this means that most of the stock exhibits a relatively homogeneous age structure, whereas a smaller proportion, especially in the new federal states, will experience a very rapid change of age structure as a result of the many new construction and upgrading measures. The bridge management strategy will thus relate primarily to the existing stock in the old federal states.

Because of the generally favourable age structure, the question of the load-carrying capacity of existing bridges is not yet all that important for the management measures to be planned, despite the fact that the maximum permissible weights and axle loads for heavy vehicles have been raised several times. Around 75 % of the bridge stock is in bridge category 60/30 or 60 and can be used without restrictions by general traffic. Only 15 % of the stock has had to be classified in category 30 or below because of recalculations and thus equipped with weight restriction signs. However, given that gross weights and axle loads are due to be raised further within the framework of European harmonization while the load-carrying reserves of the structures will decrease as a result of ageing and wear, it is likely that, in the future, more consideration will have to be given to this question, too. Appropriate evaluation and control possibilities will thus have to be an essential component of a future bridge management system. However, the considerable increase in the number of movements by vehicles carrying exceptionally large goods or heavy goods is already a serious problem today. The approval procedure for such movements has to be thoroughly revised, and the movements also have to be incorporated into a bridge management system in the future.

3 ORGANIZATION OF THE ROAD CONSTRUCTION ADMINISTRATION

In Germany, the road network is divided into four categories: federal trunk roads, Länder roads, district roads and local roads. In each case, the construction agency responsible is always a public authority; the proportion of private roads is negligible.

The public construction agency for federal trunk roads is the Federal Government. It thus has responsibility for this important part of the road network, and provides the funds required.

The Basic Law (constitution) stipulates that the federal states act as agents for the Federal Government in designing, constructing, operating and maintaining federal trunk roads. Parliament (the Bundestag) decides, within the framework of a requirement plan, which road sections are to be built and stipulates the level of funding for new construction, upgrading work, operation and maintenance in the budget. The Federal Ministry of Transport controls the implementation of the requirement plan and monitors its proper execution by the federal states.

When developing a bridge management system, this division of responsibilities has to be taken into consideration as an essential prerequisite. The system has to be designed such that the interfaces between the parties involved are organized accordingly and that the Federal Government is given the means necessary for information, control and monitoring. Another essential prerequisite for the development of a bridge management system is that a standard procedure be ensured in all 16 federal states.

4 OBJECTIVES OF SYSTEMATIC ROAD MANAGEMENT

In the future, roads will continue to be the most important mode of transport for carrying goods and passenger traffic. Germany's central location within the trans-European road arteries and the increasing importance of East-West traffic following the opening of the borders mean that the primary road network in Germany is of outstanding importance for the economy and society, from both a national and international point of view.

Preserving the functionality of this road network is thus a central task for all parties involved.

The essential objectives of systematic road management must therefore include:
– Preserving the safety and functionality of roads and their component installations;
– Complying with the level of politically, economically and technically predefined requirements;
– Evening out differences in level in order to achieve similar conditions in the network or giving priority to certain corridors;
– Employing financial, personnel and ecological resources in an economically optimized manner;
– Minimizing user costs;
– Drawing up forecasts and programmes for short-, medium- and long-term control of management;
– Interconnecting the complex requirements and procedures to form an overall model;

– Permanent feedback on the current condition and likely changes in the network in order to monitor compliance with the requirements.

5 STRUCTURE FOR A SYSTEMATIC ROAD MANAGEMENT SYSTEM

In order to achieve the objectives of a systematic road management system, tools and methods have to be provided which enable the parties involved to purposefully control the complex procedures.

In a highly simplified form, the overall model can be subdivided into seven principal subject groups:
– Data collection and evaluation
– Examination of various strategies
– Forecasting of requirements
– Translation into budget requirements
– Creation of management programmes
– Execution of work
– Success evaluation, taking stock

In this context, management systems such as pavement management, bridge management and construction site management are major tools for providing the necessary information and scenarios for the specialist groups.

6 CONCEPT FOR A FUTURE BRIDGE MANAGEMENT SYSTEM IN GERMANY

The Constitution stipulates that the major responsibilities of the Federal Government as a public construction agency are to be delegated to the federal states. In the past, different systems for the organization of bridge management have evolved in the federal states. There are two reasons for this: first, the widely varying administrative structures in the federal states; and second, different conceptual approaches and different database systems. Uniform standards have tended to be confined to technical requirements in regulations and specifications and to procedural instructions. So far, the Federal Government has controlled the management of the bridge stock only on a very small scale, and the maintenance strategies of the federal states have primarily related to the project level.

In recent years, however, the improved possibilities produced by the computerized processing of the complex connections and the increasing realization that a uniform concept is required to implement systematic road management has led to an intensification of cooperation between the Federal Government and the federal states in this sphere. This also includes the development of a bridge management system, which is to be established step by step in the form of modules. Tools and procedures that already exist are to be integrated wherever possible. This requires an in-depth analysis of the current status, and this is currently being conducted within the framework of a study.

The 1992 OECD report on "Bridge Management" [1] states that a bridge management system (BMS) should encompass all engineering and management functions that are necessary to efficiently carry out bridge operations. The system should include formal procedures for coordinating these functions and analytical tools or models to help identify bridge needs and establish priorities.

In particular, a BMS should consider the condition of the entire bridge inventory (network level) when allocating resources and establishing maintenance policies. According to [1], the prototype for a BMS should be structured as shown in Figure 3.

With due consideration given to the division of responsibilities between the Federal Government and the federal states in the sphere of federal trunk roads, a future BMS will have to satisfy two major requirements:
– It will have to enable the Federal Government to obtain not only an overview of the current condition of the structures at network level but also information on funding requirements, and to realize strategies, long-term objectives and general conditions in management practice.
– It will have to provide the federal states and authorities with recommendations for the implementation of improvements at project level which are compatible with the strategies, long-term objectives, general conditions and budgetary constraints.

The Federal Ministry of Transport has assigned the task of developing and coordinating such a BMS to the Federal Highway Research Institute (BASt), which, as a subordinate research institute, provides assistance to the Federal Government on issues relating to road transport and road construction. Its work on this project is being accompanied by a joint Federal Government/federal states working group.

The concept provides for a division of the system into a total of seven modules, which are in turn subdivided into subject groups. Some of the subject groups are interlinked both vertically and horizontally. Figure 4 shows a simplified version of the modules.

An important prerequisite for a BMS is the provision of basic data and condition data. Module I contains the necessary basic data for bridges, tunnels and other civil engineering structures, which are

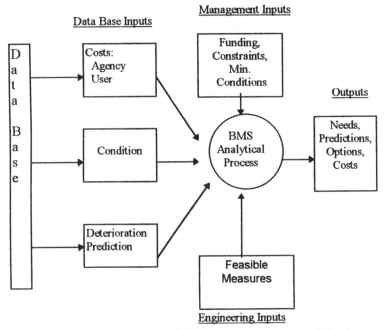

Figure 3: Fundamental structure of a bridge management system [1]

Modul I:	Modul II:	Modul III:	Modul IV:	Modul V:	Modul VI:	Modul VII:
Network level basic data	Network level condition data and evaluation	Project level analysis of defects	Rehabili-tation operation	BMS analytical process	Planning and execution	Quality securing success evaluation

Figure 4: Modules of a comprehensive bridge management system

organized as part of a comprehensive road information base. This means that recourse to other data, such as traffic data, accident data, operating data, etc., is possible at all times and a link-up with other management systems is ensured. The road information base is currently being evolved jointly by the Federal Government and federal states by means of a separate research project and adapted to a uniform standard.

Module II concerns the provision of condition data derived from the results of bridge inspections and the evaluation of the damage identified. In accordance with Standard DIN 1076 [2], bridge inspections are to be carried out every 3 years as general inspections and every 6 years as major inspections. Bridges are inspected and damage information collected and initially evaluated by specially trained bridge inspectors of the federal

states. The result of Module II is to provide a network-related condition rating and an initial calculation of funding requirements.

Module III relates to a more detailed project-related damage analysis, with experts being consulted if necessary. To this end, damage development models are to be integrated at a later date.

Module IV provides a detailed catalogue of management measures, some of which are described in more detail in regulations. Using the catalogue of measures, suitable variants can be selected and considered in each individual case.

Module V constitutes a further essential component of a BMS, although it is one that is still in its infancy. It relates to the economic evaluation of management measures within the framework of economic efficiency considerations. To this end, service life cost models and cost-benefit calculations are to be integrated at a later date. Up to now, such considerations have only been carried out in individual cases, for instance if the economic efficiency of a planned repair measure is to be evaluated in comparison with bringing forward rehabilitation or reconstruction.

Module V is also an interface between project-related and network-related approaches. Here, the results of the previous modules are combined and the maintenance measures are allocated to an urgency rating within the framework of an overall analysis. This in turn forms the basis for calculating short-, medium- and long-term funding requirements, taking into account the marginal political, economic and technical conditions. Whereas the more operational tasks of Modules I to IV are principally allocated to the lower administrative level, Module V is of primarily strategic importance and is allocated to the intermediate and higher administrative levels of the federal states.

As an important control element, Module V is also an interface between the states acting as agents for the Federal Government and the Federal Government as the public construction agency. To this end, the results from Module V are passed on to a higher management system, in which the analysis is conducted at federal level.

On the basis of the financial planning and provision of funds derived from Module V, Module VI concerns the planning and implementation of the management measures. This includes drawing up management programmes on the basis of the urgency rating determined as well as preparing and carrying out construction activities.

Finally, Module VII refers to questions of quality assurance and success evaluation.

7. FOCAL POINTS OF CURRENT DEVELOPMENT ACTIVITIES

The status of development of the individual modules varies widely at present. There are still major deficiencies in the case of, in particular, the project-related damage analysis, including damage development models (Module III), the economic efficiency considerations and the analysis at network level (Module V).

The most urgent tasks are currently to improve and expand the basic data (Module I) and to improve the provision and evaluation of the condition data (Module II). Work on developing standards for the necessary expansion of the basic data and adaptation to the updated road information base was completed in 1997, and these standards have been introduced as

the "ASB Bauwerke" [3] and are legally binding for the federal trunk road sector. Thus, in the future, the basic data of the structures can be entered directly into the database after every construction measure (repair, conversion, new construction) with the help of a newly developed data processing program. To enlarge the data stock for existing structures, it will be necessary to collect additional data. A special program is currently being prepared for this purpose.

To improve the provision and evaluation of condition data, the Technical University of Darmstadt [4] and the Federal Highway Research Institute [5] have, over the past few years, conducted research projects on behalf of the Federal Ministry of Transport. The aim was to develop a computerized system for the prioritization of management measures as a basis for an urgency rating and to achieve automatic linking with the associated costs.

A new system of automatic condition evaluation to uniform criteria eliminates existing weak spots. Within the scope of bridge inspection, the inspector evaluates each individual piece of damage with regard to its impact on traffic safety, stability and durability. On the basis of this three-phase damage evaluation, algorithms which take into account the extent of the damage and the significance of the damaged elements are used to evaluate the condition of groups of elements and of the structure as a whole.

The condition rating procedure is complemented by a planned procedure for the monetary evaluation of damage, which can also be used at any time to draw up initial cost forecasts for the maintenance requirements of the total stock of structures. In addition, based on the condition evaluation procedure, damage and condition development models are being developed for use within the scope of maintenance planning. With the condition data, condition development data and cost data then available, it will be possible to economically evaluate different measures within the scope of economic efficiency considerations. A number of research projects are being prepared for the evolution of the sub-modules.

What is important for the Federal Government, as the public construction agency, is that in addition to a bridge management system, other management systems, such as pavement management and construction site management, and improvements to the forecasting of overall requirements be developed, in order to achieve a comprehensive system for systematic road management. Such a comprehensive system is likewise currently being prepared.

8 COMPARISON WITH BRIDGE MANAGEMENT SYSTEMS IN OTHER COUNTRIES

Since the early 1980s, the individual states of the USA have used bridge management systems to support their maintenance planning. Measures to optimize maintenance planning became necessary when the extent of the damage to road bridges was realized. Around 70 % of the approximately 580,000 bridges on the US road network were constructed before 1935. The advanced age of many bridges, combined with the fact that these bridges were designed for lower volumes of traffic, plus the chemical and physical strains to which they have been subjected, have led to serious structural problems. Since it was not possible in the past to provide sufficient funds for their maintenance or replacement, around 40 % of bridge structures are today considered to be damaged or aged [6].

The first bridge management systems used by the states were not standardized regarding their structure or the tasks they were to perform and the requirements they were to meet. Since the early 1990s, the Federal Highway Administration (FHWA) has been developing a comprehensive BMS known as "PONTIS", which is now being used by many states [7].

The structure of PONTIS is characterized by conditions in the USA. First, the focal point of damage models was geared to specific structural problems. Second, in the USA, identically designed types of structure, with which good experience had been gained, were frequently used in the past [7]. This made it possible for PONTIS to consider elements and groups of elements, types of damage and deterioration processes plus necessary measures within a limited range of variation.

The German trunk road network exhibits different conditions. First, the mean age of bridge structures is significantly lower than that of bridges in the USA. Second, greater importance has been attached to bridge maintenance, at least in the old federal states, which means that the deficiencies and damage are basically less of a structural nature and relate more to traffic safety and durability problems. Moreover, almost every one of the 34,632 bridges on the German trunk road network is unique.

Thus, the development of a BMS for the German trunk road network can correspond to the US systems only in its basic features. Fundamental simplifications and generalizations as in PONTIS are not appropriate in a German BMS. Moreover, the different administrative structures have to be taken into account.

9. CONCLUSIONS

In the future, roads will continue to be the most important mode of transport for goods and passenger traffic in Germany and Europe. Preserving their functionality, especially of the primary trunk road networks, is thus of outstanding importance for the economy and society.

Almost all countries are currently developing models for systematic road management. In this context, computerized management systems are important tools for optimizing the organization of procedures and interconnections, some of which are very complex.

A major component of a future overall road management system is the development of a bridge management system, which can be used to comprehensively record and evaluate the stock of structures at the project and network levels. The particular marginal conditions regarding the structure of this stock and the legally predetermined administrative structure have to be taken into consideration as major prerequisites.

The concept for a future bridge management system in Germany provides for division into a total of seven modules, each of which is subdivided into a number of different subject groups. Development work is being carried out jointly by the Federal Government and the federal states. The current development phase is focusing on enlarging the stock of basic data and improving the recording and evaluation of conditions by means of evolved DP programs. The objective of this development phase is to achieve largely automated procedures for rating the urgency and determining the funding requirements of maintenance work.

Because of the complex nature of the task involved, international cooperation is especially important in this sphere, so as to be able to incorporate other countries' experiences into development activities and to exchange ideas on how to improve the systems. The task itself is, in principle, almost identical in all countries, even if national peculiarities have to be taken into account.

It is therefore in the interest of all parties involved to provide the public with better information on the tasks that lie ahead and to convince policymakers of the need for timely action.

REFERENCES

[1]   Bridge Management, Report prepared by an OECD Scientific Expert Group, Paris, 1992

[2]   DIN 1076, Ingenieurbauwerke im Zuge von

Straßen und Wegen. Uberwachung und Prüfung,
1983 edition

[3]   ASB Anweisung Straßen Teilsystem
Bauwerksdaten, 1997 edition

[4]   Schubert, Hitzel, Managementsystem
Brückenerhaltung 3. Stufe, Research Report,
Technical University of Darmstadt, FE No. 15.
253 R 95 H, 1997 (unpublished)

[5]   P. Haardt, Erarbeitung von Kriterien zur
Zustandserfassung und Schadensbeurteilung von
Brücken- und Ingenieurbauwerken, Research
Report, Federal Highway Research Institute, FE
No. 97 24 3/B 4, 1997 (unpublished)

[6]   D. O'Connor, W. Hyman, Bridge
Management Systems, FHWA Report
No. DP - 71 - 01 R, Washington, 1989

[7]   N.N., Report on the 1995 Scanning Review
of   European Bridge Structures, NCHRP Report
381, Transportation Research Board, Washington,
1996

Operation and Maintenance of Large Infrastructure Projects, Vincentsen & Jensen (eds)
© 1998 Taylor & Francis, ISBN 90 5410 963 7

Safety management for infrastructure projects: State of the art and future trends

H. Bohnenblust
Department for Risk and Safety Planning, Ernst Basler + Partners Ltd, Zollikon, Switzerland

H. Bossert
Department for Project Management, Ernst Basler + Partners Ltd, Zollikon, Switzerland

ABSTRACT: As introduction a brief description is given of the new Swiss Transalpine Railway Link (AlpTransit). It includes an overview of the technical concept and its special challenges, as well as information on the current planning stage.

In the second part, safety management of infrastructure projects is discussed from a general point of view, focusing on three important elements:

(1) A sound and comprehensive safety management policy,
(2) a responsible and adaptable safety management organization and
(3) tools and criteria for identifying, assessing and evaluating the different technical, financial and environmental risks based on a common safety concept.

In the third part an illustration is given of how such a safety management plan is implemented within the Swiss project AlpTransit. It is shown how the current risk management activities fit into a long-term policy, ensuring that the AlpTransit link can be operated safely and economically.

1 THE NEW SWISS TRANSALPINE RAILWAY LINK

1.1 *The AlpTransit Railway Network*

The new Swiss Transalpine Railway Link (Alp-Transit) will be an integral part of the European high-speed railway network for the transport of passengers and freight. The new railway through the Swiss Alps will allow freight to be transferred from the congested roads to rail, and at the same time linking Switzerland to the high-speed passenger transport system between Northern Europe and Italy. Upon completion, the AlpTransit network will actively contribute to reducing road-traffic-related environmental pollution in Switzerland, and undoubtedly will lead to similar improvements in neighboring countries.

The two parallel axes Gotthard and Lötschberg reflect technical and political needs as well as traffic flow (see Figure 1).

The Gotthard axis is the key part of AlpTransit and will be able to accommodate 240 freight and 60 passenger trains daily. Passenger trains will travel at speeds of about 200 to 250 km/h, and freight trains will run at 100 to 160 km/h. This will increase passenger capacity from 5 mio. passengers/year today to

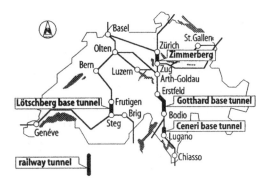

Figure 1. The Transalpine axes Gotthard and Lötschberg in Switzerland.

8 mio. passengers/year in 2020, and freight capacity from 12 mio. tons/year to 50 mio. tons/year in the same time span. Furthermore, the new railway will halve the traveling time from Zurich to Milan to only two hours.

With its apex at only 550 meters above sea level, the Gotthard axis will be the lowest of all existing and planned transalpine routes (see Figure 2).

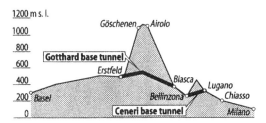

Figure 2. Longitudinal north-south profile through Switzerland.

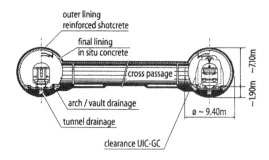

Figure 3. Cross-section through the Gotthard tunnel.

1.2 *Planning Stages*

Up to 1991, the Federal Transport Ministry carried out the initial studies for the project. In 1991 the Swiss parliament, in accordance with the inter-European contract, decided to build two new railways through the Alps. As usual for large national projects, the Swiss population had to confirm the decision in a public vote. The issue was voted on and accepted in September 1992. In 1993 the Swiss Federal Railways (SFR) were designated as owners of the new Gotthard line, from Zurich to Lugano through the Central Alps, and they were made responsible for the design, construction, and subsequent operation.

1.3 *The Gotthard Base Tunnel as the Core of the Project*

On April 12, 1995, the Swiss Federal Council approved the preliminary study for the new Gotthard base tunnel between the portals Erstfeld in the north and Bodio in the south.

The Gotthard base tunnel actually consists of two tubes, one for each travel direction (see Figure 3). Detailed studies have shown that this system is the most cost-effective, has the shortest construction time, limits the risk in geologically problematic areas to a minimum, and meets the demanding safety requirements. Two tunnel cross-over points, where the trains can move from one tunnel to the other, guarantee that capacity demands can be met, and they also ensure operational efficiency during maintenance work (see Figure 4).

For construction of the tunnel, the Swiss Federal Railways plan to invite tenders for five packages. Two will start from the portals, two from the intermediate attacks in Amsteg and Faido, and one from Sedrun through a vertical shaft with a depth of 800 m.

The shaft of Sedrun is already under construction, while work at the intermediate attacks at Amsteg and at Faido will start in spring 1999. If all goes as planned, the tunnel will be ready for operation in the year 2010.

2 SAFETY MANAGEMENT PLAN

2.1 *Why do we need a Safety Management Plan?*

The construction, operation and maintenance of large infrastructure projects involve many different issues. Safety is just one of them. However, safety touches upon most other issues and the systematic management of safety can simplify the project as a whole.

Today's large infrastructure projects are very complex systems. Large sections go through tunnels or cross bridges. Trains travel at higher speeds, traffic density on rail and road increases. New technologies are being used, and sometimes the limits of feasibility need to be adjusted. As a result new risk areas may be encountered.

At the same time, most projects face tight budgets and tough time schedules. Private investors are in the background. Identifying safety issues as early as possible and properly assigning responsibilities helps to find solutions in good time and to prevent cost overruns and delays.

In the past, large infrastructure projects were usually in the hands of government agencies. Today many of these are in the process of restructuring and privatization, whereby the keywords are increased efficiency and cost effectiveness. This can be in direct conflict with the goal of providing safe services. A representative of British Rail summarized this as follows: „After changing from a quasi-military structure to a business-oriented approach British Rail improved phenomenally but perhaps safety got side-lined". In the future cost effectiveness has to be addressed explicitly in safety issues as well.

Those in charge of large infrastructure projects need to assume overall responsibility by considering all safety issues and making sure that no aspect is left out. The primary goal on infrastructure projects is to promote economic and cultural development by providing fast, reliable and comfortable transportation. At the same time, possible negative side effects such as environmental impacts and accidental events

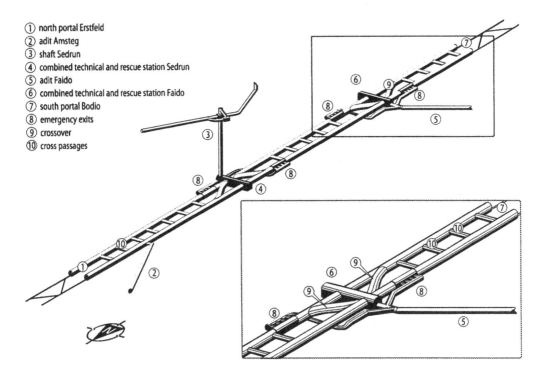

Figure 4. View of the Gotthard base tunnel.

must be considered. Since adequate safety is taken for granted, awareness of risk issues is usually lower in the initial phases of a project. In many situations, however, existing codes and regulations will not be sufficient. To deal with these situations a coherent decision making process is needed.

Furthermore, public awareness towards safety issues is increasing. This is especially true with respect to catastrophic accidents involving a large number of fatalities and injuries. Today, risk issues are no longer the preserve of the technical professionals directly involved. More and more frequently, politicians, the general public, insurance companies and others demand detailed information and open discussion of the risks involved. Effective communication with parties in- and outside the project requires a common basis: knowledge of the nature and size of risks, their relative importance in the project, their absolute importance compared to other societal risks, and knowledge of the spectrum of possible safety measures and their effectiveness.

In short, many infrastructure projects face a dilemma: how can profitability be achieved and maintained and – at the same time – an excellent safety record be retained? Decisions will have to be made where the two objectives are in conflict with each other. It then becomes essential, that the resources available for the enhancement of safety be optimally allocated, to ensure a maximum safety benefit. A comprehensive safety management plan may help to achieve this requirement.

2.2 What is a Safety Management Plan?

A safety management plan is an action-oriented decision-aiding tool. It takes a holistic view in the sense of requiring all safety issues to be evaluated by the same rules that form part of a company's overall objectives. The general characteristics of a safety management plan can be described as follows:

– It focuses on the *optimal allocation of resources*. Since tight budgets ask for the careful evaluation of all actions taken, cost effectiveness is of increasing concern. Safety measures need to be evaluated with respect to cost effectiveness as well. This does not mean saving on safety. Rather, it ensures obtaining maximum safety for the money spent.

– It allows *the linking of facts and value judgements*. Safety planning calls for knowledge of the technical and scientific facts of the processes related to a company's activities, coupled with a sound understanding of the value judgments made by the people involved and affected. Facts and value judgements cannot be considered separately.

They have to be integrated within the same framework.

- It ensures *consistent and transparent decisions.* Infrastructure systems are becoming more and more complex. Numerous decisions have to be taken, ranging from routine decisions by maintenance managers up to the strategic decision whether to allow the transportation of hazardous goods. Many of these decisions affect safety issues. The safety management plan helps to ensure that these decisions are taken in a consistent manner that complies with a company's safety targets.

- It *simplifies communication concerning safety* among specialists, as well as with the public and with authorities. Communication concerning safety is not easy because facts and value judgements are hard to keep apart. Also, there is no common language, which makes communication difficult, even for specialists. Communication with the public and with authorities is increasingly important. The risks should be known and communicated. Tools are needed to demonstrate how decisions are going to affect those risks.

- It allows *control of the effect of actions* taken. The effect of safety measures may not always be apparent. The annual accident statistics may not reveal enough evidence of improvement or deterioration in a system with little or no accidents. By using a systematic, analytical framework the effect of safety measures can be shown.

2.3 *Elements and Goals of a Safety Management Plan*

Providing safety is not a one-off action, but a continuing, dynamic process. In accordance with this a safety management plan consists not only of an analytical framework, but includes process-oriented elements as well. Figure 5 indicates these elements: the *safety policy statement*; the *organizational structure*; the *risk-based safety analysis*, and the *communication culture*.

Figure 5. The elements of a safety management plan.

A safety management plan is a tool to reach the following goals:

- Ensuring that all safety issues are treated comprehensively, individually and systematically, according to predetermined principles and criteria, and according to the requirements of the respective project phase, be it construction, operation or maintenance.

- Ensuring that the organizational arrangements are adequate for handling the safety issues according to defined responsibilities, rules and procedures during all phases of the project, and that organization is adapted to the current requirements.

- Ensuring that the information flow to parties in- and outside the project and to the general public is adequate in order to gain and maintain approval and general confidence.

2.4 *Safety Policy Statement*

The safety policy statement describes a company's vision with respect to safety. It can be compared to the environmental policy statements many companies have issued in recent years. It includes the guiding principles that direct the company and its employees while acting on safety issues. It puts the value judgements and attitudes towards safety into a set of operable rules. In that sense it acts as a means to motivate and guide all staff members. It allows top management to delegate safety decisions to the appropriate management level and to ensure that all decisions will be consistent. The safety policy statement also acts as a tool to communicate with the public and with authorities.

A safety policy statement will need to be thoroughly discussed with and finally approved by the top management. It needs to be communicated to all staff members, to authorities and to the public at large.

A short outline of possible topics to be covered in a safety policy statement is given in Table 1.

Table 1. Topics to be covered in a safety policy statement.

1. Safe transportation of passengers and freight; safety of third parties and environment; safety of staff members.
2. Safety as part of the corporate culture
3. Innovations for the improvement of safety
4. Respecting existing rules and standards
5. Knowing the risks and recognizing that „zero risk" is not achievable
6. Minimizing the probability of accidental events, as well as the damage caused by accidents
7. Striving for a balanced safety level, based on the cost effectiveness of safety measures
8. Adequate safety organization and safety controlling
9. Informing on and communicating safety issues

2.5 Organizational Structure

Many of the decisions made throughout a company's organization affect safety issues. The organizational structure is the means through which a company achieves its general targets, as well as its goals with respect to safety. A careful examination of a company's organizational structure from the viewpoint of safety will reveal today's information flow, opinion-influencing processes and decision processes with respect to safety. To achieve the goals of a safety management plan it is crucial that all parties involved understand the responsibilities with respect to safety. The following aspects have proved essential in this respect:

On the normative level the top management needs to be involved. It issues the safety policy statement and thus defining the basic rules for safety management. It is advisable to create a safety board which includes part or all of the top management. Major issues on safety should be reviewed and approved by the safety board. The board balances the views of the company against those of the public. It plays an essential role in „living" an appropriate safety culture and in communication with authorities and the public.

On the strategic level the overall direction of the safety management plan and the organizational development are defined. The strategic management deals with „the external and the future", such as comparisons to similar projects, changes in legal requirements and the definition of safety criteria. It performs planning and intelligence functions. One of its main tasks is the review and controlling of the safety analysis. The strategic management should include the responsible staff of all divisions and be headed by the safety officer. It should meet regularly as a „safety team" to discuss safety issues only.

On the operative level the line managers are responsible for implementation and control. They deal with „the inside and now". They make sure that all rules and standards are complied with, safety measures are actually put into operation and that risks are analyzed if and when necessary. In short, it is the daily business of the line managers to ensure safe operation.

2.6 Safety concept

Figure 6 indicates the basic elements of a safety concept. For each phase of the project *(definition and description of the system)* the safety concept pinpoints the existing risks in the system *(risk analysis)* and how these should be appraised *(risk criteria)*. On the basis of the risk assessment, additional safety measures will be taken if necessary, in order that the system may be designated sufficiently safe *(system is safe)*.

The procedure of the safety concept is always the same for all project phases, regardless of the type of

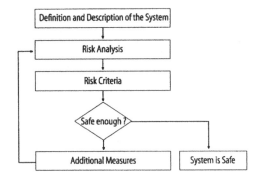

Figure 6. Elements of the safety concept.

risks or the degree of itemization in the specifications. For example, the same procedures are applied for occupational safety during the construction phase as for the subsequent safety of passengers in the operation phase.

When carrying out the risk analysis, the hazards are identified and the expected frequency and consequences are assessed. The use of quantitative methods – always adapted to the depth required – is part of the state of the art (see Figure 7).

While risk analysis deals with facts, risk appraisal concerns value judgements. These include estimated acceptance of damage and readiness to take preventive measures. Risk appraisal cannot be made objectively and must therefore be based on a decision of the parties involved.

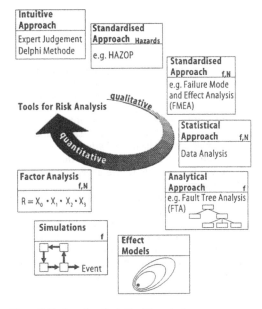

Figure 7. Tools and methods for risk analysis.

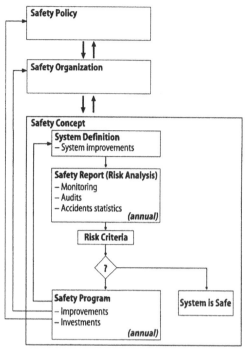

Figure 8. Risk-cost-diagram.

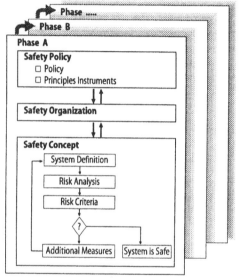

Figure 9 Process in the construction phase referring to the respective project phase.

Figure 10. Process in the operating phase with continually developing safety concept.

Since risks are recorded quantitatively in the risk analysis, it is logical for the risk appraisal also to be based on quantitative factors. Tools like fN-diagrams with acceptance areas (ALARP) and quantitative cost/effect criteria (marginal costs) are widely used. According to Figure 8 additional safety measures usually entail costs which – whether consciously or unconsciously – are included in the appraisal. Therefore it is preferable to use explicitly formulated cost effectiveness criteria.

Measures are planned in several steps, from evaluation of risk-reducing measures up to assessment of risk reduction and costs for these measures.

The safety concept is documented in the form of a report and it serves both for quality assurance purposes and as documentation and proof of planned or implemented safety measures vis-à-vis third parties (e.g. safety authority).

2.7 Safety Process

The safety management plan unites the three components of policy, organization and concept, each with scope and functions relating to the project phase in question (Figure 9). The stipulated safety policy, organization and safety concept reflect the actual status of a particular project phase.

If the project continues up to the operating phase the deployment of personnel and capacities may change, but not the basic structure of the safety management plan. As indicated in Figure 10 cycles will then run not from one phase of the project to another but, for example, as annual adaptations.

3 IMPLEMENTATION OF THE SAFETY MANAGEMENT PLAN IN THE PROJECT ALPTRANSIT GOTTHARD

3.1 Strategic Risk Issues

Risk and safety have been an issue from the very beginning of the planning process in the AlpTransit Gotthard project. At an early stage a number of *strategic risk issues* were identified, which could have a major impact on time and cost of the project or on the future operation and maintenance:

46

– Geological risks of potentially unstable rock formations;
– Tunnel system concept (single- or double-track tubes with or without service tunnel);
– Concept of underground service stations as emergency and rescue elements;
– Transportation of hazardous goods;
– Fires in passenger trains and rescue procedures for this case;
– Occupational safety and health during construction (e.g. vertical shaft of 800 m and temperatures up to 50 °C) and operation;
– Requirements for railway and signaling technology.

During the phase of the feasibility studies these issues were not included in a formal safety management plan, but investigated by groups of specialists. After starting the conceptual planning and design phase, however, a more formal safety management plan was gradually developed. The comprehensive examination and appraisal of all relevant safety issues became especially important during this phase, because the AlpTransit Gotthard project, including all safety issues, were open to public appeal.

3.2 Safety Policy Statement

A *safety policy statement* has been formulated and adopted by the top management of the AlpTransit Gotthard organization and the Swiss Federal Railways (see Table 2). This policy undergoes periodical review in order to take new developments into account and to meet the requirements of the current planning and construction phase.

Table 2. Excerpts of the safety policy statement of AlpTransit Gotthard.

⇒ *The safety policy statement is documented in a safety management plan. It deals with all risks to persons, environment and property during construction and operation*
⇒ *Safety is the result of the optimal combination of structural, technical, operational and organizational measures. However small risks as an element of any technology cannot be avoided.*
⇒ *The safety level shall correspond to those achieved on other European high performance tracks, and appropriately compare to other new railway lines in Switzerland.*
⇒ *Safety decision shall be based on explicit criteria, taking into account existing rules and regulations as well as the cost effectiveness.*
⇒ *The safety management plan shall be based on a self-adapting safety organization and auditing and review procedures.*

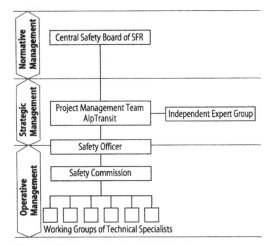

Figure 11. Safety organization of AlpTransit.

3.3 Safety Organization

At present, the *safety organization* (see Figure 11) consists of the following elements:

The *central safety board* of the Swiss Federal Railways is responsible for the normative safety management dealing with the overall safety policy and culture of SFR, safety goals and programs. It is also responsible for communicating safety issues to the public. Its members include the director generals of the main departments of SFR and the head of the safety department.

The project management team of AlpTransit Gotthard is responsible for implementing the safety policy in the project and reports to the central safety board. It has designated a *safety officer* who is in charge of the technical implementation and enforcement and thus being responsible for strategic management (e.g. risk criteria, guidelines) and the operative management (implementation of rules and attitudes). The office of the safety officer is staffed by safety specialists from SFR.

In addition, the project management of AlpTransit Gotthard has set up an *independent expert group* whose task is the critical review of all safety issues, as well as ensuring that the project corresponds to the current international state of art. This group reports directly to the project management team and consists of railway and non-railway safety experts from Switzerland and abroad.

In order to implement the safety policy on a broad level the safety officer has established the *safety commission AlpTransit Gotthard* in which the different components of the project are represented (construction, signaling, rolling stock, operation, maintenance, emergency and rescue). This commission is responsible for defining safety standards and criteria as well as coordinating the investigation of the different *specialized working groups*. These working groups deal with questions like aerodynamics and climate that have to be considered in the safety design, operational requirements during emergency situations, specifications for emergency stations, occupational safety during construction and operation etc. The safety commission and the working groups consist of specialists from SFR and external consultants.

3.4 *Current Status of the Safety concept*

The *safety concept* of AlpTransit Gotthard is based on a systematic and mostly quantitative probabilistic analysis of all risks to person and the environment. The acceptability of risk or the need for additional safety measures is defined by quantitative risk criteria for individual and collective risks. These criteria include cost effectiveness considerations.

Based on the safety concept, a safety case has been submitted to the authorities for approval of the conceptual design of the Gotthard base tunnel. The document, which is accessible to the public, describes the current state of the safety policy and the safety organization and gives an overview of relevant accident scenarios and the associated risks (see Table 3). The safety case concludes that the current conceptual design of the Gotthard base tunnel satisfies the safety requirements for persons and the environment. The Swiss Federal Transport Ministry as responsible authority, is currently reviewing the safety case.

3.5 *Future Work*

In the next step of the detailed design and tender

phase the identified key safety elements are investigated in more detail. The following questions are of prime interest:

- What are the structural and technical options for ensuring optimal conditions in the emergency stations during serious fires in passenger trains?
- What are the operational requirements for successful rescue and evacuation of trains that have stopped between emergency stations?
- What are the effects of ventilation in the case of accidents with fires or release of hazardous goods?
- Which limitations (if any) must be placed on the transportation of hazardous goods?
- What are the specifications on passenger train cars and how do they compare with international interoperability standards?

REFERENCES

1. Fachgruppe Sicherheit AlpTransit Gotthard 1995. *Sicherheitskonzept, Abschnitt Basistunnel*. Projektleitung AlpTransit Gotthard.
2. Ernst Basler + Partners Ltd. 1993. *Overall Safety Management: Concept and Recommendations for Implementation*. Øresundskonsortiet.
3. Ernst Basler + Partners Ltd. 1993. *Risikoorientierte Sicherheitsnachweise im Eisenbahnbetrieb*. Bundesministerium für Verkehr.
4. Rayner, D. 1993. *How British Rail puts a Price on Safety*. In *DYP Risk Update* 79/Oct. 1993.

Table 3. Key elements of the safety concept AlpTransit.

⇒ *Tunnel layout designed to support safety aspects*

- *Two single-track tunnels with cross passages at intervals of 325 Meters.*

- *Two emergency and rescue stations, built into the service stations with fresh air access and smoke ventilation.*

⇒ *Railway technology to minimize risks:*

- *Operation guided by traffic control center*

- *Cab signaling and automatic speed control*

- *Train radio*

- *Access limited to passenger trains with defined standards*

⇒ *Maintenance concept based on partial and temporary closure of tunnel sections*

⇒ *Emergency and rescue procedures: fire brigades permanently available*

Operation and Maintenance of Large Infrastructure Projects, Vincentsen & Jensen (eds)
© 1998 Taylor & Francis, ISBN 90 5410 963 7

Quality objectives for the operation and maintenance of bridges and tunnels

Jørn Lauridsen
Bridge Department, Danish Road Directorate, København, Denmark

Erik Stoklund Larsen
COWI Consulting Engineers and Planners AS, Lyngby, Denmark

ABSTRACT: The present article gives an overview of quality aims for the operation and maintenance of bridges and tunnels. These aims are based on the requirements of the community and are related to the following four main topics:
• Safety
• Passability
• Environment
• Service
Each topic covers several aims, and the way in which they are achieved is described. With regard to safety, the influence of inspections, inspection intervals, the technical quality of the work, installations and monitoring systems are described. Passability for road-users at all times is important. The article describes the influence of preventive measures, repair strategies, the administration of special transports and emergency plans for ensuring passability. Environmental aspects are discussed, e.g. health, noise, vibration, aesthetics, etc. Service and information for the road-users and their effect on improving passability is discussed. In general terms, it is described how quality aims influence each other and how they are linked by the need for economic optimum management of bridges and tunnels.

1 INTRODUCTION

The pressure on the road sector is increasing. During the past ten years, the volume of traffic has increased by 3 - 4% per year. The road network has been expanded, improved and maintained to keep pace with this growth, and a flexible management has contributed to maintaining traffic flow with no reduction in safety. The bridges and tunnels naturally form part of the road network.

An overriding requirement is that the safety level on the road network, including bridges and tunnels, is maintained, while the road network is operated in the economic optimum way.

Many factors influence the choice of operation and maintenance strategy for bridges and tunnels. They include the political situation, budgets, road-users, the environment, aesthetics, damage and repair methods.

It is the task of the management authority to take all these factors into account in the operation of bridges and tunnels, i.e. to try to form an overall view.

2 SAFETY

Safety is maintained by keeping the structures under observation by means of inspections and by requiring the rehabilitation works to be of high quality.

2.1 Inspections

The chief aim of inspections, as mentioned above, is to ensure that traffic safety is maintained. A further aim is to monitor changes in the condition of the structures, including the registration of defects, and to give the management authority the necessary technical and economic basis for planning improvements, cleaning and maintenance, in order that these activities are carried out in the optimum way throughout the lifetime of the structures. Inspection procedures are described in detail in the Danish Road Regulations.

2.1.1 *Making an inventory*

The purpose of making an inventory is to give the management authority an overview of the bridges and tunnels on the road network. This is done by setting up a list of structures and establishing a structured archive system containing the most important administrative, technical and economic data. The inventory should be continuously updated, as in addition to the above-mentioned data, information on important changes during the lifetime of the structure are registered.

2.1.2 *Routine inspections*

The purpose of routine inspections is to ensure that the structures fulfil their function, i.e. that no damage or other factor that increases risk has arisen. Routine inspections are of two types:
- Normal routine inspection
- Extended routine inspection

Normal routine inspection should be carried out together with the road net inspection, so that the road inspection intervals are followed, i.e. 1 - 2 times per week.

Extended routine inspection should be carried out twice a year. These inspections also include the road section under the bridge, as the inspection is used for planning and checking the routine cleaning and maintenance work. Expanded routine inspection may also be ordered because of special circumstances, e.g. a collision, heavy rainfall or high water level.

If a routine inspection reveals damage that affects traffic safety, the damage must either be remedied immediately or appropriate warning signs must be set up. For other types of damage, a report is made so that the damage can be remedied in the course of subsequent maintenance.

2.1.3 *General inspections*

The chief aim of general inspections is to provide a basis for initiating activities so that traffic safety can be maintained. A general inspection is a thorough and systematic visual inspection of all the components of a structure. General inspections should be carried out at intervals of 1 to 6 years, so that the development of damage can be monitored and remedial measures carried out at the optimum time.

The interval is adjusted to the condition of the structure and the traffic volume so that safety is maintained while the amount of inspection work is minimized. The intervals are based on 25 years of experience.

The inspections should be carried out by engineers with a knowledge of statics and the mechanisms of deterioration. This makes it possible to evaluate a structure on the spot and thereby minimize the amount of data reported (only significant damage is reported).

In a general inspection, passage data (vertical and horizontal clearances, etc.) are registered, and the structural components are evaluated on the basis of observed damage. A condition rating is then given to each component.

On the basis of component evaluations, it is noted whether a special inspection (see below) is needed, together with the method, estimated cost and time of execution of any remedial measures that are considered necessary. The general inspection thus gives the management authority an overview of the condition of the structure. The general inspection can be used for provisional budgeting and priority-ranking of major remedial measures and final priority-ranking of minor repairs. It can also be used to evaluate the cleanliness and maintenance condition.

2.1.4 *Special inspections*

The purpose of special inspections is to give the management authority a more precise basis for making decisions on the priority-ranking of major improvements, including replacements and works needed to maintain the safety standard. Special inspections are normally carried out when recommended after a general inspection, and occasionally after a routine inspection. A special inspection results in more detailed information on the structure. Special inspections are carried out by specialists (engineers) when there is a specific reason for the inspection, and are therefore not carried out at regular intervals.

There are two types of special inspection:
- Economic special inspection
- Technical special inspection

In an economic special inspection, a proposal containing several improvement strategies for the structure is prepared. For each strategy, the method, time-schedule and cost are specified. Economic special inspection should always be carried out before starting a major improvement.

A technical special inspection normally involves

investigations to determine the extent and causes of observed damage. Test specimens are usually taken from structural components, and laboratory investigations may also be carried out. Technical special inspection can also include special investigations, e.g. periodic measurements required as a result of collisions, floods, etc. An evaluation of load-carrying capacity can also form part of a technical special inspection. The technical special inspection includes the economic special inspection

2.2 *Installations and monitoring systems*

Installations and monitoring systems are mounted on certain structures in order to evaluate the condition and operation of the structure.

For example, electronic monitoring equipment (SCADA) automatically registers a traffic stop and alarms a supervision centre or the police. By means of cameras the police can identify the cause and call help and/or rescue teams.

In certain structures warning sensors are built in ; these give information on traffic flows, slippery roads, etc.

Finally, certain structures have both sensors and measuring points, so that routine measurements can give information on the technical state of the structure (settlements, moisture, chloride penetration in concrete, etc.).

Figure 1. Rehabilitation of concrete columns.

2.3 *Technical quality of work executed*

In operation and maintenance work it is important to maintain a high level of quality. It is normal in Denmark for contractors to have a quality assurance system that meets the requirements of the ISO 9000 series. The Road Directorate has participated in the preparation of product and execution standards that specify the repairs and how they are to be carried out. The Road Directorate also has guidelines for the preparation of tender documents for routine maintenance and major repairs.

The tender guidelines have been formulated on the basis of over 40 years experience in the repair of structures. These guidelines contain general specifications (AAB) which describe how repairs are to be carried out. For each repair job a special specification (SAB) is prepared which takes the specific conditions into account. These guidelines make up together with a strong quality assurance system and professional contractors, form the basis that is necessary to fulfil the aim of a very high degree of safety on the constructions.

3 PASSABILITY

High passability is necessary for safety and effectiveness of the transport effectiveness. Experience show that accidents, such as end-to-end collisions, often take place where the vehicles have piled up due to insufficient passability. These situations arise as a result of traffic diversions, i.e. narrowing of the carriageway and consequent reduction of capacity, usually in connection with road repairs.

For safety reasons it is important to maintain the maximum possible passability.

The section on traffic-economic unit-prices clearly shows that poor passability on the road network involves heavy costs for society.

To increase or maintain passability, action can be taken on several fronts:
- Preventive maintenance (e.g. filling cracks in asphalt surfacing) that can postpone major repair jobs.
- Optimum repair strategies and projects.
- Special transports are enabled to travel rapidly and safely on the road net.
- Emergency plans are prepared which take effect in the event of a traffic jam or structural failure.
- Making information available (cf. the section on service).

3.1 Preventive measures

Preventive measures are minor works that can be carried out without interfering
with the traffic. They are generally based on the principle "away with water", i.e. they include sweeping, cleaning of manholes, drains and expansion joints, spot painting of steel, repair of asphalt cracks, etc.

Routine inspection and maintenance of electrical and mechanical installations are also counted as preventive measures.

Preventive measures are usually cheap and can postpone major repairs and their associated traffic problems and heavy costs for several years. This results in greater passability and gives economic advantages in the long run, cf. the section on economy.

3.2 Repair works

Sooner or later preventive measures can no longer postpone a major repair, and one or more structural components must be repaired or replaced. When a major repair is necessary, several methods of carrying it out should be investigated.

For each proposed method the technical details must be specified; if the method affects traffic flow, a proposal for dealing with the traffic during the repair work must be included.

The proposals can vary from no action at all to replacement of the entire structure. It should be emphasized that in addition to the technical evaluation, traffic and economic evaluations should be made.

In the traffic evaluation prices are allotted to inconveniences to road-users. These prices are used in the economic evaluation.

3.3 Special transports

Special transports are those that exceed the permissible limits for overall weight, axle load or dimensions. Such transports travel on the road network every day, and it is an important aim to facilitate the work involved in permitting these transports.

Special transports are taken into account in the design of bridges and tunnels by giving them additional load-carrying capacity as well as horizontal and vertical clearances that exceed the prescribed minimum.

Danish bridges are designed for a vehicle loading of three 26-ton axles and three 13-ton axles (two 3-axled vehicles side by side). This ensures that an actual vehicle load of at least 100 tons (often considerably more) can pass the bridge without restrictions. With restrictions in the form of reduced speed, the use of specified traffic lanes and the exclusion of other traffic while the special transport is on the bridge, the permissible load can be increased by at least 25%. The cost of dimensioning bridges for these heavy loads is normally marginal and the high load-carrying capacity is a great advantage for the transport sector. It also means that main roads can be used by these vehicles and that diversions to avoid weak bridges or road stretches are unnecessary.

In principle the same applies to high vehicles, as the aim is a vertical clearance of at least 4.0 m under all bridges, corresponding to the maximum permissible vehicle height. All new bridges in Denmark have a vertical clearance of 4.5 m, corresponding to the clearance on European motorways. This makes accidents caused by the sudden braking of high vehicles due to the driver being in doubt whether there is sufficient clearance less likely. The clearance ensures that damage to the bridges due to high vehicles seldom occurs.

It is also an aim to expedite the issuing of permits for special transports, thereby assisting the transport sector. In Denmark the police deal with simple cases, while the road authority deals with extremely heavy transports. As most countries have more than one road authority it can be difficult to obtain permits for special transports because of the number of different authorities. In future, passage data will be collected in a database for carrying capacity and clearances covering the entire road network. This will facilitate the administration of special transports.

Figure 2. Special transport.

3.4 *Emergency plans*

Emergency plans should be prepared for special structures. The plan takes effect in the event of an accident, fire or structural failure.

The plan must specify how the rescue authority is to be alarmed, and how the rescue team is to reach the scene. The procedures to be specified include:

*   reaching the structure
*   reaching the location of the accident (incl. the route)
*   how the cause of the alarm is localized
*   how the traffic flow is to operate in the event of an alarm

The following measures must be described:

*   how contact with the centre is maintained
*   how the traffic installations are to be operated in an emergency
*   where access to the emergency controls is located
*   special circumstances, e.g. leakages of toxic or inflammable liquids.

4 ENVIRONMENT

In accordance with overall traffic and environmental objectives, a road authority should work towards developing the transport sector in an environmentally acceptable way. The Road Directorate will therefore:

*   In collaboration with authorities at home and abroad, with citizens and other parties involved in transport and environment, strive to advance environmentally acceptable, safe and effective solutions of transportation problems.
*   Continuously strive to prevent or minimize the negative environmental effects of the road infrastructure on people and nature while taking the given framework into account.
*   Advance cleaner technologies and a safe and healthy working environment in the production and operation processes of the road sector.
*   Take the environment into account in management decisions at all levels, and motivate and train staff to act in accordance with the Directorate's policy.
*   Communicate openly on the Directorate's environmental policy, on the environmental consequences of its activities and on the Directorate's contribution to solving environmental problems.

The Road Directorate has decided to focus on the following environmental problems in connection with bridges and tunnels:

*   Effects on landscape and cultural values
*   Pollution of soil and water
*   Consumption of raw materials and manufactured products
*   Noise
*   Visual pollution

4.1 *Effects on landscape and cultural values*

An important aim in today's planning is the protection of our natural and cultural heritage, so that future generations can enjoy it. A road will often involve a considerable change in the landscape and can affect a wide belt of the surroundings. It is therefore important to pay attention to environmental consequences at an early stage of the planning. This means that attention must be paid to the landscape and surroundings when a new road is designed. To take the many factors involved into account, EEE-investigations (Evaluation of Effects on the Environment) have become a natural part of planning and design. In EEE-investigations, the landscape, geological, hydrological, biological and culture-historical conditions are investigated in order to avoid or minimize negative effects from the new road. Furthermore, extensive aesthetic evaluations of the landscape are carried out to ensure that the project harmonizes with the surroundings as well as possible.

To reduce the large and increasing number of small mammals, birds, frogs and toads killed on the Danish roads, fauna passages are incorporated into new road projects to an increasing extent. This has been neglected until now on existing roads, but there is no doubt that fauna passages will be provided on these roads as well. A fauna passage can be a relatively wide bridge screened from the underpass road with bushes, etc., or it can be an underpass, e.g. along a stream, in which there is room for a passage. It is noteworthy that almost all valley crossings are now bridges instead of embankments. The aim here is to preserve the open valley profile and retain the development possibilities of the valley.

4.2 *Pollution of soil and water*

Since the beginning of the 80s the Road Directorate has provided motorways with closed drainage systems and special basins with oil separators for the accumulation of runoff. A certain amount of heavy

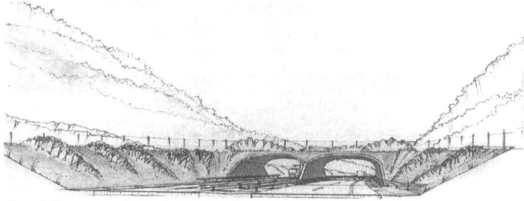

Figure 3: Fauna passage (Møller & Grønborg AS)

metals, oil, etc. separate out in the basins, so that these substances do not enter streams and watercourses or pollute the groundwater.

The basins can also help to prevent toxic substances released in road accidents from reaching the groundwater. The sediment in the basins is collected at regular intervals, and the sludge is deposited in supervised dumps. Runoff from existing bridges over water has not yet been dealt with; the runoff is usually led into the fjord, stream or sea. Accidents on these bridges in which toxic substances are released can be a danger to the environment, and remedial action will be taken in due course.

Environmental damage from road salt chiefly affects vegetation along the roads - with a risk of pollution of surface water and groundwater. Salting of roads to avoid slippery surfaces is minimized by salting at the right time in relation to frost or rain and by using the minimum amount that will do the job. In recent years there have been attempts to find a better thawing agent for use on concrete structures. Urea was used for a time as it does not give rise to corrosion in steel. However, its use was stopped because chemical compounds harmful to the environment were formed. Currently, tests are in progress with CMA (calcium-magnesium-acetate), which is a good thawing agent and non-corrosive to steel, but is expensive. The aim is to find a production method that will reduce the cost.

The Road Directorate is trying to find an alternative to herbicides for combating growth of weeds on the roads. The technical feasibility, economy and environmental consequences of various methods are under evaluation. The use of herbicides on structures has ceased. This applies especially to weeds on slopes under bridges and along wing-walls and supports.

Finally, operations in streams and watercourses (inspection, cleaning, repairs) can result in pollution and the transfer of bacteria from one stream to another. This can be a danger to e.g. fish-farms.

4.3 Consumption of raw materials and manufactured products

The road sector is a major consumer of materials for building and maintaining roads. In general, efforts are made to reserve the highest quality raw materials for uses where the technical functions are most demanding. Whenever possible, poorer quality materials that are abundant are preferred to high quality materials or materials in short supply.

Today almost all broken-up asphalt is re-used. Most of it is used in the production of new asphalt materials. Waste and residues from other sectors are also widely used, e.g. slag, fly ash, broken concrete and broken brick.

As the most common material used in bridge construction is concrete, it will be necessary to investigate:
- The use of new cements with lower CO_2 emission when produced.
- New concrete mixes with other materials.
- The possibility of reducing the consumption of materials
- Changes in operation and maintenance the above-mentioned will involve.

4.4 Noise

Road traffic is the principal source of noise pollution in Denmark. Many houses and flats are located in

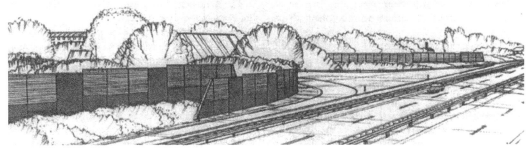

Figure 4. Noise-screens (Møller & Grønborg AS)

areas where traffic noise exceeds the level considered to be acceptable. In addition, noise can be a problem for recreative areas and nature parks; noise can disturb animals and affect the value of the area for people.

The setting up of noise-screens along roads will reduce the number of dwellings affected by road noise. Dwellings where the noise level exceeds 65 dB(A) are given the highest priority.

Bridges are often at a high level relative to the surrounding terrain, which can result in a noise problem. Setting up noise-screens on bridges gives rise to the following problems:
- Problems in the maintenance of edge-beams where noise-screens are attached.
- Road-users usually want a good field of view when passing over bridges. This difficulty can be overcome by using transparent screens.
- Bridges are often prominent in the landscape. Aesthetics must therefore be taken into account in the choice of noise-screen.

4.5 Visual pollution

The road should be considered as a whole, i.e. the road itself, its installations and immediate surroundings. A road that harmonizes with the landscape must not have its aesthetic qualities destroyed by large buildings or signboards.

Aesthetics are now an element in all phases of planning, construction, operation, maintenance and administration of the road network. Bridges on the road net can be considered as interruptions for the road-users. It is important that the bridges and their surroundings be "cleaned up" so that no drain-pipes, railings, lamp standards etc. destroy the overall picture. It is also important to preserve old bridges, which are frequently considered as having aesthetic value.

4.6 Safety and health

In Denmark, a safety and health plan must be prepared when one or more employers have more than 10 persons simultaneously on the site, or the estimated amount of work exceeds 30 working days and at least 20 persons simultaneously employed or 500 man-days or when the work involves special dangers.

All repairs on bridges are in the above-mentioned category.

The plan, which must take site conditions into account, should be in document form by the time a project is put out to tender. The plan must exist as a document before work begins, and shall include:
- Site organization, i.e. relevant data for the client, the designer, the planner, the safety coordinator and the structure of the safety organization.
- Overview drawing of the site, showing the site engineer's office, the workmen's huts, parking area, technical aids (cranes, mixing plant, cutting and bending bench, etc.), lighting, access and transport roads, storage area, rubbish containers, technical installations, first aid gear, existing installations under ground level and areas with special risks.
- A time-schedule that shows how the various works or work phases stand in relation to one another.
- A description of all factors of significance for safety and health, to ensure that each employer can carry out his task within the law on working environment.
- A description of relevant precautions, including marking of existing underground installations such as high-voltage cables, gas pipes, etc., as well as special risks on the area such as storage of toxic material, polluted soil, etc.
- General measures, e.g. who does what and when, quality requirements.
- General welfare measures.

- Risks connected with the project, e.g. the use of dangerous materials, special working methods or processes.
- Emergency plan if there is a possibility of emission of harmful substances or danger of fire, explosion, accidents, etc.
- A list of the employers involved in the project.

The plan is especially relevant for structural repairs in which it is necessary to work close to heavy traffic, at heights, or with dangerous materials. Increasingly strict limits on the working environment can be expected, so that development projects are needed. Here it is sufficient to mention robot-controlled water-chiselling, developed in order to avoid "white fingers" (vibration-induced damage to blood-vessels in the hands).

4.7 *Environmentally oriented design*

The environmental orientation of design is preventive, and aims at reducing the resource consumption of the structure and its negative effects on health and the environment in the construction, operation and demolition phases.

In Denmark a handbook in environmentally oriented design has been developed; it takes many of the factors described above into account.

All methods of environmental evaluation aim at creating the best possible overall picture of the environmental consequences of the project, and hereby make it possible for decision-makers to identify possibilities of improvement. In the so-called EEE-investigations (evaluation of effects on the environment) briefly referred to above, the environmental status before and after the project is determined. It is also important to bear in mind the so-called life-cycle evaluations, which are primarily an attempt to evaluate all the processes that take place during the lifetime of the structure, including operation, maintenance and materials.

5 SERVICE

The Road Directorate aims at providing service both for road-users and the Danish road sector. Examples are the administration of special transports and inquiries from road-users and people living near roads.

The Road Directorate also initiates development projects, prepares road regulations, and takes part in the formulation of standards. These activities benefit the entire road sector.

The following fields are included in the service functions of the Road Directorate:
- Service in connection with planned repairs via newspapers, TV, advertisements and signposting to avoid misunderstandings and prevent traffic problems.
- As bridges attract the attention of the public, the Directorate answers questions on bridges, organizes guided visits to bridges, and issues publications on bridges.
- Major bridge owners are under an obligation to provide service for smaller bridge owners. The Road Directorate has an information section that issues reports, notes, brochures, guides, etc. on all aspects of the road sector.

6 ECONOMY

Achieving the above-mentioned aims costs money and as resources are limited they must be priority-ranked. Priority-ranking is based on traffic-economic unit prices. These unit prices cover many activities, but not all. The calculations must therefore often be supplemented by separate evaluations. *The overriding aim of bridge management is to achieve economic optimum management, taking the economic interests of society as a whole into account.* It is therefore necessary to consider factors that cannot be given a price. For economic comparison the so called the present value method is used.

6.1 *The present value method*

In connection with a bridge repair it is often necessary to choose between various strategies. Shall one choose an expensive repair with a long lifetime or a cheaper repair with a short lifetime? Another problem is the time at which the repair should be carried out. Should it be done as soon as possible, can it be deferred, or can it wait until the structure is replaced? An economic calculation method that can help in such decisions - the present value method - will be described in the following.

In a present value calculation, the costs of repairs, traffic diversions, traffic noise and pollution, operation and maintenance are calculated year for year within a chosen time-horizon; the timing of each cost is based on the lifetime of each repair. The annual amounts are then discounted back to the initial year using a given discount rate. In this way the present value of each year's expenditure is obtained.

By summing the present values, a value for the strategy in question is obtained that can be compared with the corresponding value for other strategies. The strategy for which the cumulative present value is lowest is the economic optimum for the structure considered in isolation.

The cumulative present value makes it possible to compare strategies in which the costs are spread over varying periods, as all costs are converted to the initial year. The further in the future a cost falls due, the lower is the present value of that cost. This effect is proportional to the discount rate adopted. To put it simply, the present value is the amount that must be deposited in the bank today to cover a cost that will fall due at the time the repair is carried out.

The present value is calculated by

$$I_n = I/(1 + r)^n$$

where I_n is the present value of a cost I in year n
I is the cost in year n based on the chosen price level (normally the current price)
n is the number of years until the cost falls due
r is the discount rate, decided by the management authority

The present value calculation is thus carried out in fixed prices (those of the initial year) with a chosen price level and a chosen discount rate.

6.2 *Parameters for the present value*

In an economic evaluation of alternatives, the most important parameters are:
- The costs of repairs and traffic inconveniences.
- The equal-value element in the repair strategies.
- The lifetime of the structural components.
- The time-horizon of the calculations
- The residual value.
- The time at which the repair is carried out.
- The discount rate

6.2.1 *Costs*

The cost of each repair strategy within a given time-horizon is calculated, the following items being taken into account to the extent relevant:
- Repair costs, which include the costs of the contractor and the consultant.
- Client supply items and any railway costs that may be involved.

Table 1. Summary of traffic-economic unit prices.

Travel costs excl. taxes	
Passenger cars	0.83 kr/km
Lorries	1.06 kr/km
Time costs	
Passenger cars	51.17 kr/h
Lorries	149.65 kr/h
Traffic accident costs	
Per reported injury to a person	1,181,000 kr
Environmental costs	
Noise costs	34,300 kr/NLI[*]
Air pollution costs	
Passenger cars	
Local pollution	0.22 kr/km
Regional pollution	0.11 kr/km
Lorries	
Local pollution	0.65 kr/km
Regional pollution	0.32 kr/km

[*] Noise loading index. The degree to which a person feels irritated by noise.

- Costs of inconvenience to road-users and other costs to society (e.g. relaying of cables and pipes), according to table 1.
- Any operational costs that have an appreciable effect on the choice of repair strategy.

It should be noted that costs to society are not included in the budgets, but are used solely in economic comparisons of strategies.

6.2.2 *Repair strategies*

Several repair strategies should be investigated for each structure. To make an economic comparison between strategies, they must result in the same increase in the value of the structure. If a strategy involves a renovation of the structure in the form of a strengthening or extension (e.g. replacement of an old safety-barrier with a new one), the value of the improvement shall be assessed, and a similar improvement must be included in the other strategies if they are to be comparable.

6.2.3 *Lifetime*

The lifetimes of the structural components in question are estimated for each strategy. They are estimated on the basis of experience with the various methods included in the strategies. The estimated lifetime takes the expected maintenance of the component into account. Safety considerations can reduce the lifetime relative to that estimated on the basis of repair and maintenance; e.g. replacement of a functioning but obsolete safety-barrier by a new safety-barrier.

6.2.4 *Time-horizon*

The time-horizon is usually that of the repair with the longest lifetime. To make the various repair strategies comparable from the economic point of view, the same time-horizon must be used for all of them. The selected time-horizon should be long enough to make the present value of expenditure beyond the horizon insignificant. The normal time-horizon is 25 years, but can be longer if the discount rate is low.

6.2.5 *The time at which the repair is carried out*

The time at which the repair is to be carried out is determined on the basis of experience. Postponement of a repair will usually result in further damage and consequently increased repair costs later. The time of execution is thus based on the economic optimum lifetime of the structural component, and is chosen so that the present value of each repair strategy is a minimum. In fixing the times of execution of a number of related repairs, the repairs should be considered as a group in order to reduce general costs, e.g. for traffic diversions.

Budgetary limitations may require a postponement of the above-mentioned times of execution. This means that the present value of the repair can increase, as the cost of repairing an increased amount of damage and possibly also the cost of increased traffic inconvenience may outweigh the economic gain resulting from the postponement.

6.2.6 *Residual value*

A consequence of using the same time-horizon for several strategies will often be that when the horizon is reached, there will be a residual value because the lifetime has not expired. If the value declines linearly with time, it is easy to calculate the residual value. Other forms of decline are possible, e.g. a parabolic curve, corresponding to a slow decline in early years and a rapid decline later. However, this will normally have little influence on the calculation, so that a linear decline can be used in most cases.

In present value calculations, the residual value is discounted and deducted from the cumulative discounted cost.

6.3 *Sensitivity analysis*

In sensitivity analysis, the effect of changes in parameters on the present value is investigated. Typically, it is the increase in repair costs resulting from a postponement of the proposed repair times that is investigated. Postponements are generally the result of changes in budgetary allocations.

Only the parameters that have a different effect on different strategies need be investigated. Parameters that have the same effect will increase or decrease the present value of all the strategies by the same factor.

6.4 *Priority-ranking of repairs*

Budgetary constraints usually make it impossible to carry out all repairs at the economic optimum time. It is therefore necessary to priority-rank the optimal repairs.

In formulating a repair strategy it is standard practice to calculate the increase in cost that would result from a five year postponement. The cost increase is typically due to an increase in the amount of demolition work or to traffic restrictions in the postponement period.

By linear extrapolation of the costs from the optimum time over the following five years, six values are obtained for the cost (from year zero to year five).

As there are usually three strategies for each repair, there are approximately 18 possibilities. Using electronic data processing, one can select, postpone or advance the repairs in such a way that the total cost is within the budget. The process is iterative, with millions of combinations. The process selects the optimum combination of repairs and execution times within the given budget for all the bridges for which repairs are planned. The chosen solutions are not necessarily the optimum solutions for the individual bridge.

By not carrying out a repair at the optimum time, as a result of grant restrictions, damage and traffic inconvenience increase and this involves additional costs. This extra cost, the sum of the differences between the prices of the economic optimum solutions and the selected solutions, is an indication of the economic loss to society that results from budgetary allocations being insufficient to cover the cost of carrying out the optimum solutions. This calculated loss is used as an argument to obtain grants to carry out the economic optimum repairs.

7 KEY INDICATORS

The above review of quality aims for the operation and maintenance of bridges and tunnels has been overshadowed by current economic conditions. Safety on the roads reflects the world and the economy in which we live. Adequate service to road-users is impossible if allocations are insufficient for daily operation and maintenance. Lack of funds will also affect the extent to which the environment can be taken into account.

To improve the management of the limited funds available, a number of key indicators can give an idea of how things are going. Another purpose of these indicators is to enable comparisons between management authorities to be made.

The following key indicators are examples:
- Total bridge area
- Average age
- Number of structural components in acceptable condition
- Repair needs
- Repair price per condition class per m^2 bridge area
- Extent of unacceptable operational conditions (%)
- Average condition class
- Optimum allocation and consequences of reduced allocation
- Permanent traffic inconvenience costs for vehicles that meet the requirements of the traffic regulations
- Temporary traffic inconvenience costs
- Answered inquiries about structures
- Number of work accidents
- Extent of improvements to the appearance of structures
- etc.

8 CONCLUSIONS

The article gives an overview of quality aims for the operation and maintenance of bridges and tunnels. These objectives are based an the requirements of the community and are related to the four main topics, such as safety, passability, environment and service. These topics are connected by the overall economic conditions, the road authority have to take care of.

Operation optimisation – Traffic management and toll collection

Operation and Maintenance of Large Infrastructure Projects, Vincentsen & Jensen (eds)
© 1998 Taylor & Francis, ISBN 90 5410 963 7

Traffic management: State of the art and future trends

J.P.F.Visser & J.J.Klijnhout
Rijkswaterstaat, Traffic Research Centre, Department of Infrastructure and ITS, Rotterdam, Netherlands

ABSTRACT: Traffic management has become an indispensable component of the service offered by a modern road system, especially in densely populated areas and on critical links. Traffic management both promotes increased safety and optimum usage of the existing road (network) capacity and is a source of valuable information for travellers. Tools cover a wide range from lane signalling and ramp metering, to route guidance and road pricing.

Dutch MTM is presented as an example of a state-of-the-art traffic management system, itself focusing on lane control but at he same time providing data to higher-level control and information systems. As a special subject, the specific requirements of permanent cross-river connections on traffic management are discussed.

Expected trends for the future are standardisation, integration, network management, in-car systems and growing consumer influence on ITS.

1 INTRODUCTION

Modern traffic presents us with many challenges. While both traffic volume and vehicle speed potential are rising, road capacity is more or less fixed, and both safety and pollution are major problems. Traffic management can be of some help in resolving the resulting conflicts.

Traffic management exists in many forms, e.g.: urban vs. inter-urban, car vs. public transport, static vs. dynamic, demand vs. supply management. This paper focusses on dynamic traffic management on motorways, with a given demand for transportation capacity.

Possible goals of dynamic traffic management include:
- reducing trip times (or at least: making them more predictable);
- increasing safety;
- reducing emissions;
- influencing the modal split.

As a side-benefit, data collected for dynamic traffic management may be used for research and policy-making purposes.

In the next section an overview will be presented of dynamic traffic management tools in the given context. In section 3 some examples will be presented from the Dutch Motorway Traffic Management system (MTM). Section 4 treats special considerations relating to bridges and tunnels. The last section of the paper encompasses some expected future trends.

2 A TRAFFIC MANAGEMENT TOOLBOX

There is no such thing as a standardised traffic management framework or architecture yet, though some try to create one. In my opinion traffic managers are still artisans with a toolbox of systems that are not necessarily related, the appliance of which is very much dependent on the specific situation.

The following classification of dynamic traffic management systems - or actually: subsystems - was roughly adopted from Van Leusden (1992) and Middelham (1995).

2.1 Monitoring systems

Knowing the current state of traffic on a stretch of road is absolutely vital to any traffic management system. Usually this is achieved by means of induction loop detectors in or below the pavement. Loop detectors have the advantage of being cheap and rather reliable for automatically measuring large streams of traffic. Among the disadvantages are their literally being "in the way" (making them a nuisance as far as maintenance is concerned) and the limitations they put onto the types of data to be collected (just speed, length, number of cars).

Alternative traffic monitoring devices, like camera's and microwave detectors, have their limitations too, and often the cost of these devices is a bit prohibitive. Therefore they are to be used instead of, or combined with, loop detectors only in specific situations.

The Netherlands are installing a Monitoring network covering all motorways and some other important roads. The technology used is that of the induction loop detector. Most of this will be operational at the end of this year (1998).

Transponders are a relatively new development. In theory, they offer almost umlimited possibilities for both vehicle detection and information from the road-side to the vehicle.

Apart from traffic itself, other data may be important to traffic management, their importance often depending on location. Among these are weather data, predominantly about wind, fog and ice. These data should generally be collected through road-side measuring devices.

2.2 Road-side based systems

Road-side based systems are (a.o.) lane signalling and ramp metering.

A lane signalling system serves to influence traffic whilst driving on a stretch of road (motorway). Among its possible functions are:
- congestion & incident warning;
- weather & road conditions warning;
- closing lanes (e.g. in case of road works);
- speed reduction;
- signalling temporary traffic rules.

Lane closure, supported by signalling system

In the Netherlands, most of these functions are performed by the Motorway Traffic Management system, which is to be discussed in section 3. An evaluation study by the Delft University of Technology showed an overall reduction of traffic accidents by about a fourth, an increase of traffic throughput during tailbacks of 4,5 % and a smoother merging at temporary bottlenecks during road works (Remeijn 1995). On the motorways equiped with signalling, that is. Nowadays (1998) this is about 700 km in both directions.

Ramp metering: it turns out that in very heavy traffic, about 2 km downstream from any influx of additional vehicles, a stop-and-go pattern or downright congestion may arise, causing shockwaves upstream. This can be prevented by controlling the influx (from a ramp), keeping it low under critical circumstances. This strategy - usually implemented through traffic lights - may result in a 3 to 5 % increase in road capacity and a reduction of average trip times.

Ramp metering is rather common in the United States. In Europe its application is still sporadic. In my country, for example, it is operational at 4 locations only. Firstly because application makes sense from a traffic system point of view only at a small percentage of all ramps. Secondly it appears to move traffic problems from the national road network to local roads, so getting support from city councils is a non-trivial issue.

2.3 Incident management

Incident management means a smart way of handling incidents (e.g. accidents) in order to clear the road as quickly as possible. Reliable detection mechanisms help identify incidents quickly and accurately. Equipment (e.g. towing truck) should be on hand near bottlenecks with a high incident frequency. And, most importantly, emergency services should be well instructed and cooperative. The latter may require e.g. towing service personnel to take over some police work if necessary; it is therefore by no means a simple measure to implement.

Incident management has a large potential for increasing road and network performance though - in the year 2010 a country-wide reorganisation of incident management in the Netherlands will result in an estimated 15 % increase of traffic efficiency in overloaded situations (Van Leusden 1992).

2.4 Vehicle control

The car industry is developing several devices for computerized control of vehicle behaviour. Probably this year (1998) several luxury car manufacturers will introduce "intelligent cruise control": a subsystem that will keep the car at a safe distance from its predecessor, while at the same time trying to keep a preset speed.

In 1997 a stretch of automated highway was demonstrated in San Diego, California - just for study purposes, of course. In July of this year (1998), the first European demonstration will take place at the A11 motorway near Alphen a.d. Rijn, The Netherlands.

2.5 Dedicated lanes

Dedicated lanes include reserved lanes for public transport vehicles, heavy good vehicles, high occupancy vehicles. Since these can be permanently assigned, these kinds of dedicated lanes do not really fit into the picture of "dynamic" traffic management.

Related, more truly dynamic concepts are:
- tidal flow lanes (reverse direction in rush hours);
- passing restrictions for trucks during rush hours (effectively creating a dynamically dedicated lane);
- use of hard shoulder as an extra lane during rush hours;
- pay-lanes: extra lanes parallel to the motorway, reserved for paying (mostly business) traffic, presumably only used during congestion etc.

2.6 Route information

Route information concerns one or several route(s): e.g. congestion, accidents, weather conditions, expected delays, special messages, advice on alternative routes. Route information is presented through in-vehicle systems or Variable Message Signs.

Of the in-vehicle systems, the ubiquitous radio traffic broadcasts are best-known. A variation on this is the Radio Data System - Traffic Message Channel, in which the driver is only alerted for messages he needs and receives them in his or her own language. This has been evaluated in the so-called Rhine corridor pilot; a production system is now under development.

In the future, in-car systems may link radio-transmitted traffic messages with static map data in order to advise the driver of the fastest route.

Variable Message Signs informing drivers about routes are well-known in Europe, e.g. in parts of Germany and in the Paris area. In the Netherlands ± 30 VMS's for route information are operational, and their number is rapidly increasing. As I know most about these VMS's in my own country - where, by the way, they are called Dynamic Route Information Panels (DRIPs) - I will limit myself to these.

We distinguish between (1) route information and (2) route choice DRIPs. Route information DRIPs inform only about the route one is on. For most people on the route, there is no feasible

alternative but to stay on the route and bear the associated inconveniences. Still being informed as such is highly valued by many drivers: e.g. if you know you are going to be late for a business appointment, you can make a phone call to have it rescheduled. Route choice DRIPs inform about different routes, offering many drivers an alternative if the desired route is congested or even blocked.

Dynamic Route Information Panel

Even with route information DRIPs (in urban areas), typically only a small percentage of the drivers (± 10 %) actually changes his or her route because of a message on the sign (e.g. Middelham 1995). For many people the alternative route simply is not feasible. Still, even a small percentage of traffic changing routes can have considerable impact on congestion.

DRIPs are located - or are going to be located - along the main arteries of the motorway network, especially:
1. near or within the largest conurbations;
2. between conurbations, at corridors from the Randstad mainports to neigbouring countries;
3. at those same corridors, near the national borders.
By the year 2000 all of this will probably be operational.

2.7 Travel information

Adequate information can promote and stimulate the use of public transport, e.g. a sign along the motorway showing the travel time to a nearby city by train, without the risk of congestion. Travel information is a demand-management tool, so I will not go into it any further.

2.8 Road pricing

Road pricing is another demand-management tool and therefore really outside of this paper's scope. Moreover, it is treated in detail in Mr Blythe's contribution to this symposium.

3 THE DUTCH MOTORWAY TRAFFIC MANAGEMENT SYSTEM (MTM)

As an example of a traffic management system, the Dutch Motorway Traffic Management system is discussed. The system has been operational for over 15 years and the scope of the system is restricted tot lane control. Still it is very much up-to-date, thanks to unremitting maintenance and development; it is also the main source of dynamic traffic data for other Advanced Traffic Management ~ and ~ Traveller Information Systems (ATMS, ATIS).

MTM was recently adopted by the Swedish National Road Authority (Vägverket) for traffic management on motorways in the Stockholm, Göteborg and Malmö areas.

3.1 The basic system concept

The visible part of the MTM system consists of overhead matrix signals on gantries. The legends used in the matrix signals are a limited number of speed indications, deflection arrows, a red cross, an end-of-limits sign, flashers and - optionally - a green arrow.

Next to each gantry is an outstation, collecting data from the loop detectors and communicating traffic throughput and average speed to a traffic control centre. From the traffic control centre commands are sent to the outstations, and the outstations in turn control the matrix signal's lamps.

In some cases, the outstation also controls peripherals like flexible signposts, and it may communicate with and take additional commands from e.g. an adjacent bridge management system.

Experiments have shown the optimal distance between the gantries to be about 700 to 800 meters, enabling the driver to have a good view of the signals on the next gantry, and not be distracted by signals on gantries further downstream. Near carriageway-splits and intersections distances of 300 to 400 meters between gantries are sometime preferable, though. Loop detectors are situated near each gantry; if gantries are more than 500 meters apart, another series of loop detectors is placed in between, connected to the first outstation upstream.

I will now focus on some of the system's functions, illustrating the broad extent of traffic management, even on a single road stretch.

3.2 Automatic incident detection

One of the primary functions of the MTM system is called automatic incident detection (AID). If the average traffic speed drops below a preset threshold - usually 35 km/h - an incident is presumed and the matrix signals on the appropriate gantry show a 50 km/h speed limit. Also, this speed limit is introduced by "70"-legends on the next gantry upstream, and in some cases "90"-legends on the second gantry upstream. The legends are accompanied by flashers, in order to attract special attention for the danger of sudden slow-downs.

The actual algorithm is a bit more complicated, including exponential smoothing and hysteresis, targeted at the prevention of instability. E.g.: the "50"-legend is only removed after the average speed has been over a different preset threshold, usually 50 km/h, for at least two minutes. This in turn generates an unwanted side-effect when combined with over-compliant driver behaviour: our Swedish colleagues report that sometimes it takes a lot of time for a tailback to disappear because drivers adhere to the indicated speed (even while it is just a recommended speed in Sweden) if the cause of the incident is already gone - a solution is being studied.

AID operates entirely unattended.

3.3 Lane closures

The other primary function of MTM is to support lane closures. A lane closure is always preceded by deflection arrows.

A traffic control centre operator indicates the road-section(-s) and lane(-s) to be closed. The MTM-system autonomously calculates the preceding deflection arrows - sometimes this yields a multiple-stage pattern - and, if applicable, combines this with the other legends that the matrix signals already display. It is also possible for an operator to prepare a measure off-line (e.g. when planning road constructions), and even to put multiple measures into simultaneous operation.

3.4 Dynamic speed limits

In special circumstances (e.g. slippery roads, storm, ice, rush hour) the traffic control centre can demand drivers to adhere to a certain maximum speed by having this speed shown on matrix signals.

If a storm- or slippery-road detector is connected to the outstation, it can also have a speed limit displayed automatically.

3.5 Fog detection

Fog detection is a special case. Some years ago a dramatic fog-accident occurred near the Dutch-Belgian border. After that we installed an experimental fog-detection system at this stretch of the A16 motorway, which is very fog-prone indeed. The fog-detection system is not actually part of MTM, but is closely linked to it, e.g. it uses the MTM's outstation and matrix signals. So far this combination has proven very effective (Remeijn 1992).

Following recent fog-accidents in Belgium, that does not have a system like MTM, some interest was aroused in our systems (Willems 1997). Also, after a special public hearing the United States' National Transportation Safety Board recommended to follow the Dutch practice in fog-prone areas (NTSB 1992).

3.6 *Data acquisition*

The MTM-system's outstations are not only connected to a traffic control centre, but many of them also provide information to a nation-wide traffic monitoring system. This system, mentioned in section 2.1, in turn provides data for route guidance and other traffic information systems.

Finally, the outstations can be used to acquire detailed information about traffic, notably speeds and intervals of inidividual vehicles. These data are to be used for traffic research purposes or to help in the analysis of (large) accidents.

3.7 *Safety considerations*

Safety was the single most important consideration when designing the MTM system, and it still is. This is reflected in many characteristics, a.o.:

- automated operation: wherever possible, the system operates autonomously, i.e. without demand for operator intervention, ruling out many human mistakes. Still, if necessary the operator can always intervene and overrule the results of automated actions;
- rule-based calculation of combined measures: the control centre checks the compatibility of multiple measures (lane closure, operator-commanded speed limits) and AID. Illegal combinations are filtered and reported to the operator, so drivers will not be confronted with dangerous combinations (like deflection arrows on the left and the right lane, both pointing to the middle lane). In the outstation a similar check is made for the combination of commands from the control centre and from other sources;
- internal diagnostics: the outstation is continuously monitoring both its own operations and that of the detectors, the lamps in the matrix signals and communications with the traffic control centre;
- graceful degradation, e.g.: if the contact between traffic control centre and outstation is lost, the outstation switches to local mode: it will retain the legend commanded by the control centre and

run its own AID program on top of that, sing data from two or three downstream series of detection loops it's connected with. Also, all outstations are fitted with a battery, keeping traffic management operations going for a few hours in case of an external power failure.

4 BRIDGES AND TUNNELS

Since the occasion of this symposium is the realisation of the Storebaelt connection, I must pay some special attention to bridges and tunnels.

A characteristic common to most bridges and tunnels is that the link they form in the road network is much used and there are no alternatives nearby. Because of cost considerations there usually is little or no shoulder, so any stalled vehicle immediately becomes both a safety hazard and a bottleneck. To make things worse, many bridges and cannot easily be accessed from the outside, which may affect emergency services. Finally, they are relatively expensive and vulnerable to chemical and fire hazards.

Therefore they qualify for the highest degree of traffic management. Which implies automatic detection and monitoring with detectors placed at 150 meters distance from each other - shorter distances did not yield faster or better incident response in our tests. Video coverage is also advisable, for verification purposes.

Bridges require special systems, like this one for wind warning purposes

The cost of these systems are small compared to the rest of the infrastructure. In a proper design the systems can accommodate the specific tunnel (pumps) and bridge (wind) monitoring systems. Furthermore each, tunnels and bridges, require their own specific services. For tunnels, these are related to handling high vehicles, obstruction and contraflow. Bridges, at least in Northern European countries, require wind warning and ice detection systems.

4.1 *Handling high vehicles*

Protecting tunnels against too high vehicles is a must. Although various automated monitoring systems are on the market, most either generate too many false alarms or are prone to missing out on a vehicle, both of which are not acceptable in most cases.

A specially developed detector though, consisting of metal triangles hanging side by side over the road, accurately detects any high vehicle and identifies the lane in which it travels. Traffic is then stopped automatically using traffic lights as well as barriers. Via loudspeakers or with the help of a vehicle on site the culprit is guided towards a (special) off-ramp.

Using physical means to detect high vehicles involves a certain risk: last year two drunken party-goers were decapitated in The Netherlands, sticking their heads out of a bus just when it was passing a somewhat rigid height detector. Careful design may prevent this.

4.2 *Dealing with obstruction*

Any obstruction of traffic should lead to a set of slow down measures upstream. Since tunnels are very vulnerable in case of accidents, the slowing down is preferably moved to a location outside (before) the tunnel itself. The following describes a strategy to achieve this.

Once the obstruction is detected, the signal immediately preceding the obstruction will show a low maximum (or advisory) speed, e.g. "50" (km/h). The other legends in the tunnel, upstream from the obstruction, will show a slightly higher speed. e.g. "70", as will the legends immediately before (outside) the tunnel. After a short interval, during which traffic in the tunnel has had some time to adapt to the "70" all "70"-s in the tunnel are changed into "50"-s. As a result, all slowing down of traffic is entirely moved to a location outside the tunnel.

4.3 *Contraflow management*

Tunnel maintenance requires tunnel bores to be closed from time to time. In those cases, contraflow operations are used. These can be arranged with the provisions an MTM-like system offers. In this case matrix signals, variable message signs and barriers are used. Instead of free merging, traffic lights can be used to meter traffic from a multilane access carriageway into a single lane. This arrangement offers a higher capacity than simple merging.

4.4 *Wind warning*

Strong winds can cause a severe hazard on high bridges. Wind monitoring systems should measure both wind-velocity and winds gust forces. A simple algorithm can be used to define what forces are still acceptable under given circumstances. Relevant circumstances are those that determine the friction between cars and the road surface. These are the amount of ice, snow or water on that surface. The first category to suffer is the car with a caravan, the next the high trucks. Restrictions require arrangements at upstream decision points in the form of variable message signs.

4.5 *Ice detection*

The temperature of the surface of a bridge is often lower than that of the surrounding land. As bridges should be kept open to traffic as much as possible, automatic sprinkler installations are used to spray salt water against ice or glazed frost on the bridge.

5 FUTURE TRENDS

I will now present some personal expectations concerning the future of traffic management. They do not necessarily reflect the opinion of the Netherlands' Ministry of Transport.

5.1 *Standardisation*

In my opinion, the state of affairs in the ATMS / ATIS industry today is comparable to that of the computer industry in the 1970's. A few big international players, many smaller national or regional industries; every supplier has its own set of standards (or architecture). On the government side, every road authority has its own set of requirements, so systems are often tailor-made.

All of this is going to change. Like the IBM PC in 1981 established a standard - however debatable - sooner or later we are going to see widely accepted standards in the traffic industry too - for all I know NTCIP is one likely candidate. Once a sufficiently wide range of standards has been accepted, both producing and buying traffic management systems will have become whole new ballgames. On the supplier side I expect a separation of hardware and software manufacturing. Specialised hardware manufacturers will compete with low prices, software manufacturers with the most advanced software. It do not rule out that 80 %

of the current traffic industry's names may disappear in the process.

Standards will - after a few years - yield lower prices and faster development of new functions.

5.2 *Integration*

Another important future trend is integration, both functional and lateral.

Functional integration: many traffic management systems have been developed separately over the years, and this is now hindering both day-to-day operations and change. For example: any one of the traffic control centres in the Netherlands is forced to use up to 18 different operator interfaces, belonging to as many applications: traffic management ~, traffic information ~, and related systems. This is crazy and integration is a *conditio sine qua non* before any new developments can be merged with the existing applications.

Lateral integration: when motorway traffic management was new, stretches of road to be managed by these systems were more or less isolated. Nowadays it is becoming ever more common that entire motorways or even motorways network are controlled by traffic management systems. Besides, traffic management on motorways and in cities are influencing each other to an increasing degree. Lateral integration, i.e. integration of systems in adjacent areas, is therefore a necessity of the near future.

A modern traffic management centre

Integration is more than just pasting a new user interface onto existing systems. It will demand major rework or even new architectures, to be based preferably on international standards.

5.3 *Higher-level traffic management*

Both standardisation and integration will, if performed properly, on one hand alleviate the job of both operator and system designer, on the other hand create new possibilities to manage traffic on a network level.

This requires a shift in our thinking, away from the manipulation of objects (like bridges, or lamps) and toward the management of transport networks. It requires highly-educated personnel, like traffic professionals instead of retrained service mechanics. Last but not least it demands for well-advised control strategies.

5.4 *In-car systems*

Judging from the number of contributions about this subject to traffic management congresses, the emergence of in-car systems is the single most important trend for now and the near future. These include in-car information systems (e.g. route guidance, freight logistics), in-car control systems (e.g. intelligent cruise control) and systems with both in-car and road-based components (e.g. automated highways, electronic toll collection systems).

I will not go into this trend any further. For more information, refer to e.g. the proceedings of ITS world congresses (e.g. ITS 1997).

5.5 *Consumer influence*

Up to this day what traffic management looks like is mostly determined by (1) government agencies and (2) available technology. The role of consumers, the drivers for whom it is all supposed to be there, is - to say the least - marginal. This too is going to change, for two reasons:
1. with the advent of in-car systems the consumer, originally a user of "free" traffic management, is going to be a paying customer;
2. electronic toll collection and pay-lanes are going to be employed more widely for regulating traffic, and to keep economic centres accessible at a charge.

He who pays the piper, calls the tune. If drivers are to pay for certain services, they will demand influence on what that service is going to be like. I'm not sure where that will get us, but have no doubt the effects will be significant. I suppose this is something to reflect on.

REFERENCES

ITS 1997. *Proceedings of the 4th World Congress on Intelligent Transport Systems*. Berlin.

Middelham, F. 1995. *State of the art in dynamic traffic management in the Netherlands*. Rotterdam: Rijkswaterstaat AVV.

Ministry of Transport, Public Works and Water Management 1994. *Towards a better use and less congestion*. The Hague.

NTSB 1992. *Proceedings of a special public hearing on fog accidents on limited access highways* (NTSB/RP-92/01; PB92-917001). Washington D.C.: National Transportation Safety Board.

Remeijn, H. 1992. *The Dutch fog-detection and -warning project*. Rotterdam: Rijkswaterstaat AVV.

Remeijn, H. 1995. *The Dutch motorway control system, 13 years of evolution*. Rotterdam: Rijkswaterstaat AVV.

Van Leusden, G. & M. Coëmet 1992. *Dynamic Traffic management in the Netherlands*. Rotterdam: Rijkswaterstaat AVV.

Willems, J. Mist-signaleringssysteem als oplossing voor file-ongevallen?. *Verkeersspecialist, december 1997*: 8-11. Diegem (B): Kluwer.

Operation and Maintenance of Large Infrastructure Projects, Vincentsen & Jensen (eds)
© 1998 Taylor & Francis, ISBN 90 5410 963 7

Traffic management: Urban/regional ITS in the Stockholm Region

Alf Peterson
Vägverket, Region Stockholm, Sweden

ABSTRACT: This paper presents a relatively short-term ITS-scheme in generic terms, that is what is going on the global arena. It also does describe how it influence the program currently being implemented and planned to be it at the Stockholm Region of the Swedish National Road Administration (SNRA) in co-operation with other parties, particularly Stockholm Municipality, The Stockholm Public Transport Company and the Police.

A description is given of the overall tasks involved in the scheme, how these will be achieved and the step-by step developments. The methodology for achieving a well functioning Traffic Management Centre with different organisations will presented as well as the effects of what has been implemented prior to 1997 as well the plans until the middle of the year 2000 are also presented.

1. BACKGROUND ON GLOBAL LEVEL

1.1 *Definitions*

Terms used	Definitions
Road Traffic Informatics, RTI	Information Technology applicable for the area of Transport, Road and Traffic
ITS	Intelligent Transport Systems
ITS America	Intelligent Transportation Society of America
ERTICO	European Road Transport Telematics Implementation Co-ordination Organisation
VERTIS	Vehicle, Road and Traffic Intelligence Society, Japan
DRIVE	Dedicated Road Infrastructure for Vehicle safety in Europe. It is a European program for research within Road Traffic Informatics
Application area	Road Traffic Informatics is divided in many areas and many ways. Examples are Traffic Control, Traffic Information, Generic pay-systems etc.

1.2 *Intelligent Transportation Systems is global*

The background for Intelligent Transportation Systems is the use of modern information technology, IT, in the field of Transportation and Traffic Management. This is particularly obvious since transportation and traffic causes global problems and that

the society is not able to construct new roads as the needs for more communication increases, we do not have the space or money for the very spacious motorway links. However there is still space for some missing essential links in the transportation system, which we can see in Denmark and the new huge bridge links or in Stockholm where there still is a need for missing links.

The global problems are visible in the sense that the pollution of the countries, continents due to traffic and transportation causes unhealthy and death. Further on it causes accidents and killed persons of significant sizes and in addition an enormous loss of time due to inefficiencies in transportation systems, lack of co-ordination between different transportation means, as Light Rail Train, Public Transport Bus-systems, private cars etc. Not the least there should also be a more efficient way to plan for lesser transportation needs when we urbanise our cities and metropolis. Generic said the problems are the same around the world even if the magnitudes can be different.

IT-technology has been one answer the reduce the magnitude of the problems and even replacing some extensive construction in regions or countries. It has became a major factor of solutions for the future. Today there is a global development, testing and implementation within Intelligent Transportation Systems going on. Many different partners participate, researchers from traffic researcher insti- tutes/universities, from car makers, telecommunica- tion and data industries, established industry in the traffic field, responsible road and traffic authorities from both city and country levels etc.

1.3 *The International Perspective*

1.3.1 Tendencies

An expression of the global view, open attitudes is that there are yearly a world congress focusing on Road Traffic Informatics. Three major organisations are responsible:

- ERTICO
- ITS America
- VERTIS

In this way Europe, USA and Japan leads the work together and share the concept of working in devel- oping, testing, evaluations and standardisation.

Today there is a practical implementation going on. There has been a shift from visions to moderate realism. Responsible Road Authorities on national and local levels are planning in a strategic way, try- ing to find suitable organisations and co-operation between the partners responsible for transport and traffic. Besides the systems shall demonstrates its qualities.

1.3.2 Co-operation

Co-operation between partners is a key-word for the implementation of Road Traffic Informatics systems, e.g. different authorities, institutes (universities), or- ganisations and industries. Nowadays a new dimen- sions has arisen. We share information, participate in

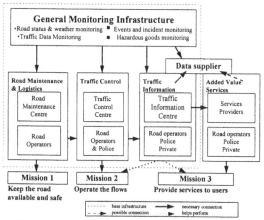

A Base structure and Base Organisation Concept

the same projects and try to transfer information, knowledge and even technologies between cities in different countries in an efficient way. Further on there is a co-operation between different public or- ganisations and public authorities, that is Public Pri- vate Partnership. Examples of this is between police and road authorities or between road authorities and commercial organisations (private or semiprivate).

1.3.3 Politics in the field of Traffic and Road Traffic Informatics

Today Road Traffic Informatics and its applications have become one of many measures or tools for Road Authorities. It should be one of the means to achieve the goals our politicians have stated for im- proved environment, traffic safety and mobility in- fluenced by traffic.

Many of the applications can or should be used by road authorities and/or police to improve the service for the road users. Other should require political considerations or decisions, like road pricing, speed control or supervision of the drivers and thirdly some are for car makers. Of course there is an com- bination of forces between the partners.

1.3.4 Application Areas

There is a consensus in how to involve or integrate Intelligent Transportation Systems in the daily oper- ating work with the partners. The figure below is an example from Europe for the main European road network and the Traffic Management on these roads.

The applications focuse are Traffic Control, Traf- fic Information Systems, Public Transportation and Parking Guidance/Management.
Within this structure different ITS- applications can be put in like Traffic Control and Traffic Informa- tion applications. There is also a strong need for co- operation to cover the total transportation need, that is between private vehicles and public transport.

1.3.5 Standardisation

An important prerequisite for the implementation nationally or for countries with exchanging informa- tion needs or traffic passing the borders is that there should be an standard for the systems to be used. Otherwise there can be a lot on non-compatible sys- tems.

For this purpose there is a European standardisa- tion going on, via CEN and its Technical Committee TC278, Road Transport and Traffic Telematics. Within this frame 14 different Working-Groups for specific topics are under progress. Within the inter-

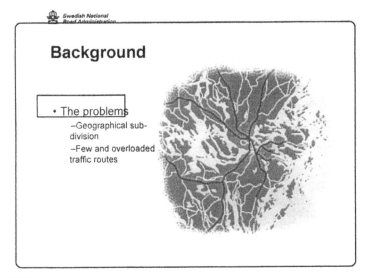

Background

Swedish National
Road Administration

- The problems
  - Geographical sub-division
  - Few and overloaded traffic routes

itv-berl.ppt

national frame the TC278 co-operate with ISO TC 204 Transport Information and Control systems.

2. WHAT'S GOING ON IN THE STOCKHOLM REGION?

2.1 Background

Stockholm is the capital of Sweden, and as such is the seat of many different public authorities and administrative offices as well as the home of diversified industries and commercial enterprises, all of which require a functional transportation infrastructure. Greater Stockholm, with a total population of approximately 1.5 million inhabitants, is comprised of several municipalities.

Stockholm stretches across numerous islands. This means that there is a limited number of communication links, (bridges for roads and railways and underground facilities) to transport the traffic that is required to move in a north-south direction through the city. This has resulted in a transport system that is highly vulnerable, specially in the event of accidents or other incidents. Three major European Highways, E4, E18 and E20, in addition to several other major roads run close by the Stockholm city centre. In addition, different public transport routes as well as railway tracks run into the very heart of the city. Although approximately 70% of all passenger transport occurs on the various public transport services, there is still an immense traffic volume on the roads.

2.2 The tasks

The overall tasks are:

- to make major roads and thoroughfares attractive enough from a safety, environmental, and accessibility point of view so as to decrease the traffic through the inner city and on the secondary road network, at schools, residential areas etc.
- to make the public transport network attractive enough that it can maintain its current level of serving 70% of the passenger transportation needs during rush hour traffic and as well maintain the 50% of passengers on daily average to city centre
- to develop an incident management system in consensus with those parties both involved in and responsible for traffic in order to ensure the swift handling of different types of incidents

2.3 Program for ITS in the Stockholm Region

This program was initiated by the Department for Transport and Communication as a request for the Stockholm Region and is sent back for political considerations.

This program aims to make the ITS possibilities visible, to have the different partners to co-operate, describe requests for investments, develop a comprehensive view on transportation where the partners for public and private transport co-operate and to spread information and knowledge.

Traffic Management Centre Stockholm

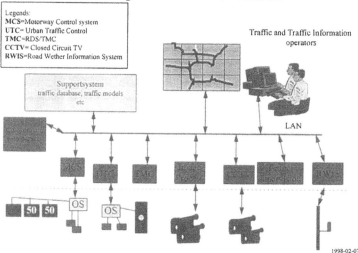

Figure 2 Traffic Management Centre in Stockholm

2.4 *Methods*

To accomplish and fulfil the tasks specified in the foregoing, the operations and activities should be implemented and evaluated in a structured step-by-step order. This should be done at the existing Traffic Management Centre level, TMC, and out along the roads as subsystems, and in connection with public transport services.

The functional organisation of the TMC is according to the figure below and is performed in a Top-down approach methodology.

2.5 *Major Functions*

2.5.1 Monitoring Traffic

One prerequisite for knowing or interpreting present traffic conditions is a system for monitoring traffic. This will mainly be based on fetching data from existing subsystem, such as UTC, MCS and probes and in an elaborated form so it can be handled by traffic models. This is under test in a development platform/area for part of Stockholm (incl. state and city roads).

2.5.2 Managing Traffic

In order to manage traffic, tools are needed to interpret the information received, to know the effect of incidents, to make recommendations possible etc. We are involved in extensive European projects for the testing, verification and evaluation of dynamic

Traffic Models operating in a real-time mode. The first chosen model to test is within Project Quartet plus and the 5T model from Torino. This is an "real-time model" including observer- and controller function together with dynamic OD-matrixes and assignment functionality's.

Along roads and streets, the efficiency of such tools as Traffic Signal Systems, UTC, is being enhanced to shorten delays, reduce incident or accident risks, and to give priority in traffic to public transport vehicles etc. according to the Swedish LHOVRA-strategy. These tools will be built in and enhanced in a successive way using optimization techniques to adapt to variations in traffic flows and with functions to automatically identify building up of queues on links. They also send data in real-time to the above mentioned Traffic Model.

Furthermore, there is a scheme for extending the existing Motorway Signalling Systems (MCS) to provide motorists with information and traffic warnings on the most occupied parts of the motorways in the greater Stockholm area. The basic functions are from the Automatic Incident Detection, AID-algorithms and lane-use algorithms. These plans includes some more 40 km MCS-systems in the Stockholm area within the next 5 years. In addition Variable Directional Signs, VDS, and Variable Message Signs, VMS, will be used to balance and to re-route traffic and inform motorist along the road and street network in the urban and suburban areas when necessary according to agreed criteria's between the partners.

2.5.3 Disseminating Traffic Information

Another tool in operation today is the co-operation with different Radio channels, public as well as commercial. This dissemination of traffic information via different media is under a plan to be enhanced with focus on its quality.

The meaning of dissemination of traffic information is to provide those using the transportation system with a basis for choosing the optimal means of transport prior to undertaking their journey, pre-trip planning. This means that different suppliers of traffic and transport information plan to share the same data bases and spread the gathered information to Radio, on internet or "info kiosk" or to a service provider for a more commercial approach.

Information will be transmitted from the roadside via VMS and received in-vehicle via Radioinformation, such as RDS/TMC. The basic recommendations are deduced automatically from the Traffic Models and from manual interventions in case of e.g. reported accidents/incidents or road works.

RDS/TMC is operational in a sharp mode since September 1997 and will give a more comprehensive information as time goes on and the co-operation with partners will be extended and improved.

2.5.4 National Road Data Base

Generic for the whole of Sweden a special national road data base, NVDB is under development. The aim for this to have one digitised road database common for all roads and streets in Sweden. Parallel with this work is a test for how to add on a digitised road data base for operational purpose where all objects are included. This work covers the test area for the development platform mentioned earlier.

2.6 *Step by step development*

Previous experience has taught us that this type of work must be done step-by-step. We have also understood the importance of testing and of co-operating with the other parties involved, both in a laboratory environment and for subsequent implementation in real situations.

This is valid

- at an organisational level in co-ordination with the systems implemented
- for building up competence and expertise within the organisations and in the areas concerned, particularly with respect to traffic management

and dynamic traffic models

- for influencing motorists or end-users
- for the gradual installation of different technical systems under progress, such as the motorway information system, MCS, (already partially in use), the improved UTC-systems including optimisation techniques as well as intelligent bus priority systems, information systems for bus passengers both at terminals, bus stops and inside buses

In this process, working together at the European level is very important as is performing evaluations, comparing results, making modifications that lead to new evaluations in an on-going interactive process.

According to the scheme, a completely new, technologically sophisticated Traffic Management Centre is to be opened in the year 2002.

2.7 *Co-operation*

During the course of our work, we have become increasingly aware of the need for agreements between road managers, the police, the fire brigade etc. to minimise the effects of an accident or incident of any kind.

These agreements covers with the City of Stockholm:

- exchange of traffic data such as flow, speed etc. for traffic models, follow-ups etc.
- exchange of information on road works, accidents that have occurred, planned public events etc. which affect the traffic on the major road network
- the road network / routes designated for traffic in the event of severe accidents or specific types of incidents
- the handling of incidents so that each party knows his exact role and the line-of-command, i.e., who is in charge on the spot, who has the authority to take immediate action, what must be done, how this is to be done, who is responsible for contacting the Traffic Management Centres so that information can be disseminated via variable message signs, re-routing signs, radio broadcasts etc.

2.8 *Results so far*

There has been different studies of the effect of using ITS in Greater Stockholm area.

2.8.1 Generic results

The most extensive investigations has been evaluations for the greater Stockholm Region level of different scenarios for two time horizons. The first one evaluation included partly year 2000 (references scenario), partly year 2010. This was performed during 1996 before the planned automatic toll debiting system was cancelled. After the cancellation of that a new scenario was evaluated and for the reference year and year 2005

I will present some results from the latter one evaluation for the years 2000 and 2005.

The scenario contains measures which influence Traffic Control, Public Transport, Trip-planning and Route Guidance. The scenarios involves the different Road Authorities in the region as well as The Public Transport company (Stockholm Lokaltrafik) and the plans for implementation for respectively partners.

The first one results are that the number of private trips will increase with up to 6% and public trips with 2-3% during the time period.

This means that the existing by-pass, The Essinge link, will increase the requirements for trips which may lead to a complete oversaturation during the rush period as you can se below in the map.

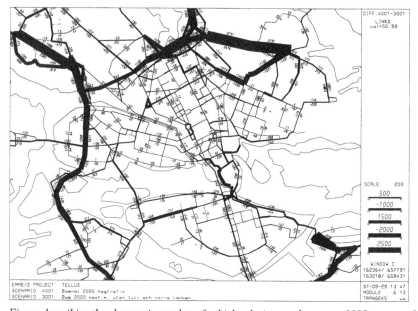

Figure describing the changes in number of vehicles during maxhour year 2005 compared with year 2000.

The calculated effects and costs for the scenario 2005 is illustrated in the table below

	Journey time	Traffic safety	Environment	Vehicle operating costs	System Costs
	Mphour	injured	NO_x (ton)	MSek	Msek
Traffic Control	-12	-180	-40	-20	55
Public Transport (travel information)	1.6	0	-40	-20	25
Trip planning	-2.2	0	0	0	10
Route Guidance	-7	-40	-110	-40	150
SUM	-23	-220	-190	-80	240

Table : Effects and costs for scenario 2005

Below is described the results in terms of a pay-off or Cost/Effect factor.

Cost/Effect	Journey time C/E	Traffic Safety C/E	Environment C/E
Traffic Control	7	0.5	2.3
Public Transport	23	>1000	0.9
Trip planning	9	>1000	>1000
Route Guidance	33	6	2.1
SUM	16	1.9	2.0

Note. Infrastructure inv.   72   1.3   0.8 sek/hour   Msek/injured
                                sek/kg NO$_x$

Table Cost/Effect year 2005

Comments

Journey-time
Cost effectiveness for all applications for the 2005 scenario from journey time point of view is approximately 16 sek/hour. Compared to investment in infrastructure this result is essential lower. This means that ITS is beneficial to implement from journey time point of view.

Among those applications mentioned specially traffic control and trip-planning has indicated high benefits. Information for Public Transport and Route Guidance are still more cost-effective than infrastructure.

Traffic safety
The cost-effectiveness from Traffic safety point of view is not that good for ITS-applications as a whole compared to investments in road infrastructure. Among those applications it is only traffic control which has a good cost-effect with 0.5 Msek per injured person.

Environment
From Environment point of view all applications involved in the scenario 2005 demonstrates a high benefit compared with traditional road investments. The cost/effect factor is 2.0 sek/kg NO$_x$ compared to 0.8 sek/kg NO$_x$ for traditional road infrastructure.

If focus on one of the applications Information in Public Transport is of special value.

2.8.2 Follow up of implemented Systems

MCS-system
Follow ups of the MCS-system after one year of operation shows a more harmonised speed along this specific stretch of the motorway, basically as a result of the AID-mechanism. This means lesser stop and go conditions and even lesser deviations from the average speed and besides a lower average speed and peak hour and as such it confirms results from studies in Netherlands.

So far it is to early to present or predict traffic safety results in terms of occurred accidents.

Attitudes has been surveyed in the form of a questionnaire that was planned and administered during the autumn of 1997. The preliminary results are as a whole positive. The majority of the road users (8 of 10) are generally positive to the system. The majority feel that they have time to see the most common symbols while driving (i.e. the speed recommended and the arrow pointing out the lane to be used). Of those having completed the form 25% stated that they had experienced problematic driving situations in the following four areas: the information system, the traffic situation, other drivers' behaviour and their own driving situation.

RDS/TMC
Another follow up is from the very first stage of the RDS/TMC application in connection with an in-vehicle device (the Volvo Dynaguide) in some 100 vehicles partly in Gothenburg, partly in Stockholm. At that time there were not that much of dynamic traffic information available from the road network, more of type reported accidents, road works, weather. The evaluation considered different indicators, such as *"the expected benefit"* (of different types of information, accidents, parking, queuing, road weather information etc.). From interviews of road users *"type of available information"* (frequency of different information), further on *"benefits"* were investigated. Traffic safety aspects were also taken into account in the interviews.

The overall result is that the drivers regard such a tool as valuable (3,5 in a scale of 1-6). The application influenced the drivers both with respect to *how to make the route choice* and *the way of driving*.

A new follow up will be done in May-June this year with the same test device and test crew as mentioned above. This time results from the real time data from traffic models and detectors will be included in the evaluation.

Ramp-metering

A third follow up is from a Ramp-Metering trial following the algorithm from Rijkswaterstaat. The evaluated effects also confirms result from Netherlands and as well the Danish test results. The pay back time is approximately one year as a function of lesser delays and higher average speeds during rush hours on the motorway. A new installation is under consideration in the Stockholm area.

2.9 *Conclusions*

ITS as being implemented so far in the Stockholm Road Traffic Management Scheme has proven to be successful to date and is running according to plans. The SNRA and the city of Stockholm are planning for a more extensive use of ITS-application in Stockholm.

Partial evaluations have been made for some of the ITS applications such as the RDS/TMC traffic information, the Motorway Signalling System, ramp metering, etc.

Operation and Maintenance of Large Infrastructure Projects, Vincentsen & Jensen (eds)
© *1998 Taylor & Francis, ISBN 90 5410 963 7*

Traffic management of highways: State-of-the-art and future trends

F. Bolte
Bundesanstalt für Strassenwesen (Federal Highway Research Institute), Bergisch Gladbach, Germany

ABSTRACT: Traffic Management (TM) has a long tradition with very positive results especially on motorways as measure to avoid congestion, to improve safety in case of tail-backs and to improve the exploitation of road network capacity. Variable message sign systems and traffic broadcasts as main instruments have been tested, fine-tuned and standardized as backbone of TM. The integration of these instruments with shared data bases yield considerable synergetic effects by mutual complementation. Emerging new telematics systems with private sector actors require reconsideration and definition of roles. Frame conditions for Public-Private Partnership (PPP) as set by the „Economic Forum for Transport Telematics" provide clear grounds for the deployment of new services in Germany.

1 INTRODUCTION

In Germany, motorways form the main backbone of road traffic. With an extension of 11 000 kilometers they comprise only 2 to 3 percent of the road network, but carry about 1/3 of the annual mileage. The geographic position of Germany in Central-Europe consequently brings a lot of international through traffic on German motorways. The traffic loads, measured as Average Daily Traffic (ADT), have increased nearly constantly over the last years (Fig. 1). The break-down of the so-called „Iron Curtain" has created additional traffic especially in east-west and west-east direction.

The motorway network has been constantly increased over the last years; it is, however, not possible to cope with the growth of traffic by a corresponding enlargement of the network. This consequently leads to more and more congestion with all negative results such as accidents, delays, economic losses and environmental impacts. Uptil now, the use of the motorways in Germany is not subject to additional payments by the users, except for heavy goods vehicles above 12 t. As the motorway network is rather dense not only long distance traffic but also medium and short distance trips such as commuter traveling is performed on the motorways. The main peak times are therefore morning and evenig hours and additionally weekend and holyday traffic. Another cause for traffic problems beside high traffic loads at peak hours are roadworks and accidents.

2 SCOPE

Traffic management has a long history in Germany. Scopes of traffic management are

- improvement of traffic safety
- exploitation of available road capacity
- reduction of environmental impacts
- better traffic economy

A number of strategies have been work out and tested in the past. The first approach is to try to avoid traffic overloads. This is done by early information about expected high traffic loads and by recommending alternative routes which offer sufficient spare capacity. Another means to achieve this is to harmonise traffic flow by traffic dependent speed regulations and specific overtaking prohibitions for heavy goods traffic.

In case that overloads cannot be avoided and thus congestion occurs the next step is to warn drivers early in advance so that they can easily adopt their speeds to prevailing conditions and avoid rear-end accidents.

In case of difficult manoeuvering actions at junctions specific dynamic lane allocations and speed regulations can help to provide better driving conditions.

3 SYSTEM

The main instruments to influence traffic flow are traffic broadcasts and variable message sign (VMS) systems (Fig. 2 and 3).

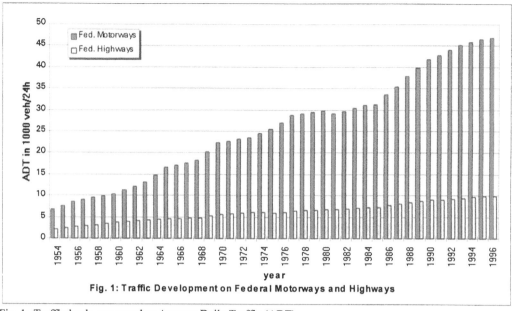

Fig. 1: Traffic Development on Federal Motorways and Highways

Fig. 1: Traffic loads, measured as Average Daily Traffic (ADT)

Fig. 2: Variable message sign (VMS) system:
Stretch Control

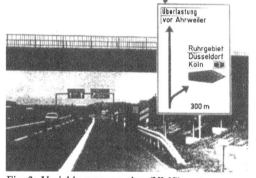

Fig. 3: Variable message sign (VMS) system:
Network Control

First programmes with traffic broadcasts were launched in the middle of the nineteen-sixties. Traffic broadcast was an instrument which used information collected by police patrols as basis for traffic messages. The operation of these traffic broadcasting services was done according to rules and guidelines as agreed by the participating partners (Fed. Ministry of Transport, Laender Road Traffic Authorities, Police, Braodcasters, Fig. 4).

The detection of incidents depended on the availability of police patrols on the scene on the motorway. A full coveridge of all problems occurring could never be guaranteed. It should,

however, be stated that these early services were highly appreciated by drivers and broadcasters.

While traffic messages in principle could cover the whole motorway network, VMS systems were installed at specific strategic points of the motorway. There are three different types of VMS systems:
- stretch control systems
- variable direction sign systems
- lane control systems at junctions

First pilots with VMS systems normally concentrated on specific problems occurring, such as congestion or fog. In order to achieve a high reliability and availability of systems, it was the

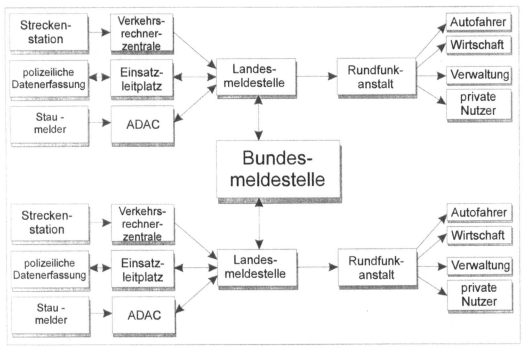

Systemvernetzung RDS - TMC - Verkehrswarndienst

Fig. 4: RDS-TMC Architecture, Actors: Police, Road Authorities, „Jam Boosters" (AdAC etc.), Broadcasters, Users, Industry

policy right from the beginning to have systems beeing operated automatically as far as possible. This was achieved with the help of suitable infrastructure, such as automatic traffic flow measurements, visibility detectors, rain detectors etc. These data are normally automatically analysed. The result of this analysis leads to the decision of taking any specific preprogrammed measures such as reduction of speed limits, rerouting or dynamic lane allocation.

The experience with these early pilots was very positive. In order to apply traffic management systems to a great extent on the motorways, suitable standards have been worked out covering all aspects to be governed. These led to a comprehensive set of prescriptions, standards and guidelines (Fig. 5).

The procurement of these guidelines had the positive side effect, that all interfaces within traffic management systems follow open specifications allowing to purchase systems and system elements in fair competition on the market. A modular configuration of systems allowed different producers to supply their products.

4 DEPLOYMENT

The Federal Ministry of Transport allocates about 100 Mio DM per year for the installation of traffic management systems. Meanwhile most of the critical parts of the German motorway network are equipped.

5 RESULTS

Accident experience with traffic management systems is generally very positive. Their results depend mainly on the quality of control. Drivers are willing to follow dynamic prescriptions und regulatons as long as they understand the reasons behind them. With well-controlled systems accident rates and accident severeness were reduced by 25 - 30 per cent each which means that accident costs rates are more or less cut by 50 per cent. Drivers are willing to follow diversion route recommendations knowing that they can avoid congestion and time losses.

The applicability of systems and strategies, of course, depends very much on the time and occasion of the occurence of incidents. During daily peak hour congestion is more or less „normal"; there is little hope to reduce congestion by diverting traffic. As most of the commuters are well aquainted with the situation the number of accidents is normally very low. Spare capacities on the available network or in the secondary network cannot be expected.

This is very different in weekend and seasonal peak traffic, where a great deal of motorists are not aquainted with the situation and the neighbouring network. In these cases early information can be used to avoid overlaod of roads by informing local people to avoid motorways. This can best be achieved by using traffic broadcasts.

It is the policy of the Ministry of Transport to make all data of collectively operating traffic mangemant systems available also for the production of traffic messages for broadcasts. The co-ordination of traffic information sources generally yields positive results for traffic management: On one hand, the basis for traffic messages is considerably enlarged. On the other hand the operation of VMS systems and traffic information messages can be co-ordinated and thus synergetic results be optained. This strategy opens new perspectives for traffic management. The co-ordinated application allows for a „task distribution" between VMS and traffic broadcasts. Variable messages signs are generally used to display signs as foreseen in Vienna Convention. This limits the amount of information to be conveyed to drivers; and it is the philosophy of Ministry of Transport to dispense as far as possible with text information as symbols are much more adequate exspecially in international traffic. Traffic broadcasts are a good means to convey information supporting prescriptions on VMS.

The coordinated use of traffic data for the creation of traffic broadcasts has caused a heavy increase of traffic messages. This leads to some negative phenomena:

- the management of messages on the information chain between road authorities, police and broadcasters had to face difficult problems due to the high amount of messages
- broadcasters were reluctant to interrupt programmes or to devote to much transmission time to broadcast traffic news. Therefore they normally cut off a number of information items in order to save transmission time; that means that information is not transmitted though it is available.
- Road users were „showered" with a high amount of messages, most of which were not relevant for them.

A way out of this dilemma was the introduction of the so-called „automized traffic broadcasts service" using a continuous data channel between all participating partners and the Radio Data System (RDS) for the transmission of messages to receivers.

Technical Specifications for Roadside Outstations (TLS)
Specifications for Variable Message Signs (RWVZ)
Specifications for Variable Message Sign Systems (RWVA)
Guidelines for Incident Detection
Guidelines for Traffic Control Models
CEN TC 226 WG3.1 Standards for VMS (under development)
Technical Delivery Specifications for Variable Message Signs (TL-WVZ) (under development)

Fig. 5: Main Guidelines for Federal Motorway and Highway Traffic Control Systems

The idea to create RDS Traffic Message Channel (RDS-TMC) for conveying traffic messages was born in 1984; only since late 1997 most parts of Germany and major parts of Europe are covered with RDS-TMC transmisson.

It would be a long story to report all problems overcome since the development of the first idea and the final deployment of the service. This story, however, is also a good example for the emerging new understanding of roles between public authorities and privates acting together along a chain for the sake of better traffic information. Major technical difficulties had to be overcome; the main obstacles, however, lay in the field of organisation, responsibility and co-operation. As all partners depended with their necessary investments on the willingnes of others to do investments on their part of the chaine a typical „chicken-and-egg problem" existed which had to be solved. Major progress was achieved when the actors succeeded in pulling all parties together around a „round table". Today, in Germany all actors are convinced that it is for their own advantage to use RDS/TMC.

6 NEW EMERGING TELEMATIC SERVICES

Considerable research funds have been invested into the development of electronics to be applied in traffic. One part was spent in developing the above mentioned traffic management services. Other funds were used to develope telematics tools which allow individual drivers (consumers) to profit from.

Industry has discovered that traffic information, route guidance, fleet managements, vehicle movement monitoring and other traffic related applications have a high market potential in the future.

This realisation of some of these systems caused some nervous reactions exspecially on the side of public authorities. The action of private operators touched their traditional area of responsibility. Very soon, the political actors discovered that conflicts between industry and private actors as service providers on one side and public authorities on the other side should be avoided with the objective to enable and support developments and not to block market access for emerging new telematics technologies. It is evident that the market requires such services and ,additionally, the introduction of these services has effects on employment rates.

In Germany, the so-called „Economic Forum on Transport Telematics" was created under the personal chairmanship of Matthias Wissmann, Federal Minister of Transport. This forum unites representatives of all concerned economic sectors plus public authorities with the objective to create frame conditions for the application of privately operated telematics services. A number of goales have already been reached: Frame conditions for access to public data have been elaborated; a general understanding of guidelines and rules for the application of „Human-Machine- Interfaces (HMI)" within cars has been agreed upon.

It is the policy of the German Ministry of Transport to remove obstacles for the deployment of telematics services on the market as much als possible and to interfere into their operation only as little as necessary. This leaves much responsibility and especially all ecnomic risks to private actors on the market. No guarantee is given by authorities for the success of service operation.

This attitude to use public and private partnership will lead to a new understanding of public authorities' role in traffic. The operation of private services does not take away any responsibility as far as safety is concerned from public actors; services going beyond the procument of safety, however, are a matter for private actors.

It is hoped that the frame conditions as set by the Economic Forum on Transport Telematics will be sufficient and strong enough to enable private-public partnership.

7 CONCLUSION

Traditional collectively operating traffic management systems using VMS and traffic broadcasts have yielded positive results in terms of accident reduction, stabilisation of trafficc flow and energy savings. Strict standardisation of functionalities, interfaces, architecture and subsystems allows systems and componenets to be purchased on an open market. Operational co-ordination brings additional synergy effects. Migration towards RDS-TMC transmission included the installation of new information channels between the participating partners. In Germany, RDS-TMC broadcasts cover the whole country.

Representatives from policy, economy and authorities, co-operating in the so-called „Economic Forum on Transport Telematics" have achieved to set up frame conditions for the development and establishment of emerging telematics services. These new services will be operated by the private sector in their own economic responsibility and without any economic guarantees from authorities.

REFERENCES

Economic Forum on Transport Telematics. Minutes of meetings. *Bundesministerium für Verkehr, Bonn, 1986 - 1987*

Bolte, F. 1997. Launching RDS-TMC Services in Europe, The German Situation, Development towards a Nationwide Service. *Proceedings of the 4th World Congress on Intelligent Transport Systems 21-24 October, 1997, Berlin. European Commission 1997*

EU-Projects FORCE/ECORTIS. *Deliverables and Internal Reports. Brüssel 1996-1998*

Bolte, F. 1997. Verkehrsmanagement - Ein Beispiel für die Zusammenarbeit von öffentlicher Hand und privaten Dienstleistungsanbietern (Traffic Management - an Example of Co-operation between Public Authorities and Private Service Providers). *Report given at 3rd AdAC/bast Symposium „Safe Driving in Europe", Baden-Baden, 11-12 June 1997*

Operation and Maintenance of Large Infrastructure Projects, Vincentsen & Jensen (eds)
© 1998 Taylor & Francis, ISBN 90 5410 963 7

Electronic tolling in Europe: State of the art and future trends

P.T. Blythe
Transport Operations Research Group, University of Newcastle upon Tyne, UK

ABSTRACT: The paper reviews the options and state of the art in electronic fee collection systems for tolling and road-use pricing.

1 INTRODUCTION

The trend in transport policy, in many parts of the world and in particularly in Europe, is increasingly towards the recovery of construction, operation and maintenance costs of new roads by the use of tolls or road-use charges. Indeed, in some cases, these charges have been extended to the existing "free" road stock. In addition to this, is an re-emergence on the political agenda of many governments and city-authorities of some form of variable road-use-pricing to address the management traffic demand.

To effectively implement the above policies it is desirable to introduce an efficient charging mechanism which is able to levy the tolls and road-use charges automatically from drivers, i.e. without the need for the drivers to perform any action (other than those associated with normal driving activities). Moreover the system should enable the collection of these charges at normal highway speeds and without the need for the separation of lanes as is the requirement with conventional toll collection facilities.

The requirement for an automatic mechanism for charging is borne out by the fact that it is deemed infeasible and unworkable, in many locations, to implement manual means of fee collection where traffic could be segregated into lanes in order that

drivers may stop their vehicles and pay a fee either, manually to an operator, or by inserting coins, cash or a card into a collecting machine. Clearly manual collection would require the building of collection plazas which are costly both to build and operate, and more crucially require a substantial land-area for the site. Such manual collection-plazas may only be built when an actual new road is conceived, land purchased and constructed as a purpose-built toll facility. It is generally not practical to 'retro-fit' a toll plaza to an existing road - in urban areas this would be wholly unacceptable on grounds of land-use, the creation of additional congestion, traffic queues and other dis-benefits, such as noise and air pollution and the inflexibility of the charging regime that can be employed. Moreover, purpose-built toll roads generally have a limited number of entry and exit points whilst "free roads" usually are not so restricted - thus creating an additional difficulty when introducing urban road charging. In order to tackle these issues more effectively it has been proposed that the charging of a road-use fee. The introduction of these charges may have a restraining effect on the traffic demand as well as the obvious attraction of raising relatively large amounts of capital which may then be put back into improving the transport infrastructure, supporting public transport and generally offering alternatives to travel by private car. To implement the policy of charging motorists efficiently the use of conventional stop and pay plazas is unattractive thus some form of non-stop automatic charging of road users must be considered.

Electronic non-stop means of collecting road-use charges or determining access rights for the driver using small communications devices in the vehicle have been trialled and implemented in Europe and elsewhere, however the performance of such systems is currently not-fully proven for large-scale implementation, however with major new installations, such as, the Storebælt bridge and the Melbourne City Link, advanced systems for toll collection which meet the new European standard are nor being deployed. Nevertheless at the moment these are the exception to the rule and importantly in the urban environment where road-pricing is desired (rather than toll-collection on inter-urban motorways) is yet to be proven.

This paper will review the types of systems and techniques can be deployed to collect road-user fees automatically. This includes all options from manual to advanced electronic systems. Where appropriate examples of these systems are cited.

2. CHARGING OPTIONS

2.1 *Parameters on which to base the charge*

The basic premise of road pricing is to charge a driver an appropriate economic fee for the drivers use of the road-space. This fee may be fixed or based solely on vehicle class (as in the case of an open toll) or may vary based upon pertinent parameters, such as:

- time of day
- class of vehicle
- prevailing level of congestion
- environmental factors
- purpose of journey
- high/low vehicle occupancy (HOV)
- other access rights

Furthermore, the payment may be in terms of a transaction at the point of passage, the purchase of a pre-paid permit, vignette or licence or the control of access based upon some other set of parameters (which may not necessarily be financial). Such payment of fees or control of vehicle access may be achieved by either manually or by some semi-automated or fully automatic means. For completeness, each is considered here:

2.2 *Manual "stop and pay"*

Manual collection methods vary in many ways, depending upon the characteristics of the road. However, the overriding requirement for manual collection is that the vehicle driver must stop the car, open a car window (or door) and either hand over cash or a card or insert either of these into a machine. These plazas are common across Europe for the collection of road tolls - no actual road pricing scheme employs such methods although arguably the Oslo and Bergen toll rings in Norway could be regarded as road-pricing installations.

Manual toll collection usually requires that a large toll plaza is built which divides the free-flowing multi-lane road into a number of single lanes. Each lane is serviced by a toll booth which either houses an operator who collects toll payments manually or the equipment (such as a card reader or coin accepting basket) which the driver may use to pay his toll. Generally, the rule for the design of plazas is that there should be at least three toll booths to service each one lane of traffic leading into the toll-plaza - clearly an non-viable option for road-use pricing in urban areas due to the size of plaza required and the high-volumes of traffic that could be expected in morning and evening 'peaks'. Figure 1 shows a 4 lane toll plaza servicing a 2-lane low-flow road in Normandy, France

Figure 1: Typical Toll Plaza Layout

Thus, a four-lane dual carriageway will require typically 12 toll booths to service the traffic efficiently. Naturally, on roads with low flows, this number may be reduced. However, it is necessary to compare the benefits of reducing the number of toll-lanes (thus the land required and the number of operators employed) with the costs of queuing, noise and air pollution which is the consequence of not being able to service the traffic efficiently and quickly at the toll plaza. Moreover, the physical security of storing and moving a large amount of coins and paper-money can cause some logistical problems. At the Mersey River Crossings in the late 1980's for example, somewhere in the region of 15% of revenue was stolen systematically by operators and about half a ton of coins were required to be moved daily.

The enforcement of manual toll systems, in general, relies on the use of a barrier which is not opened until confirmation (by the operator or the collecting machine) that the correct toll charge has been paid. In many cases, these systems are augmented by vehicle detectors (to count the vehicles passing through the lane) and by some form of vehicle classification (to distinguish different classes of vehicle which pay different tolls). Classification is usually based upon axle-counters and/or vehicle height-measuring equipment. Where a barrier is not used, a video camera may be employed. However, this practice is not very common due to the extra cost with little benefit over the barrier (as the vehicles are expected to stop anyway). Another option is to use a flashing light and alarm on the toll booth which attracts the attention of any police vehicles at the toll plaza that a vehicle has violated the system (such an approach is used extensively on the US Turnpike network) however this is not a particularly workable deterrent for congestion pricing where very high vehicle flows can be expected.

Considering the different types of manual system that exist:

Manned toll booths

Manual toll collection using an operator to collect money is probably still the most widely used method of collecting tolls. An operator is situated in a toll booth servicing one lane of traffic. These booths must be air-conditioned and heated for the comfort of the operator. It is generally also necessary to employ some simple auditing systems, such as counting the vehicles passing through the lanes and more commonly now providing a paper receipt for each transaction.

He or she collects coins, cash, tokens or paper tickets from all the drivers passing through the lane. In many cases, where only the correct toll may be paid (i.e. no change given) or where pre-paid tokens or paper tickets (vouchers) are used the transaction itself takes only a few seconds.

However, if the transaction requires that change is given or a paper receipt provided for professional drivers then this process takes longer.

In general, an experienced operator can achieve three hundred or more transactions per hour.

Automatic Coin Receptors

To replace the need for a manned toll booth, automatic coin machines are widely used at many toll collection installations. These devices are generally able to accept pre-paid tokens (if used) and coins. Most of these machines use a basket or hopper, into which the drivers throw coins/tokens. These are generally read and validated within a few seconds and the barrier raised (or some other indication of a correct toll charge given to the driver). Optionally, the driver can press a button requesting a receipt to be printed. These basket/hopper arrangements are regarded as quite an efficient way to pay tolls and are now quite reliable and environmentally robust (usually, they contain a heater/cooler to ensure that they can operate in all conditions). Moreover, the sophistication of the coin validation unit enables the machine to reject foreign currency and other objects thrown into the hopper. Figure 2 shows a combined stop and pay coin hopper, card reader and read-only tag reader on a highway near Lyon, France.

Figure 2: Coin, Card and Tag Payment Booth

For regular users of a toll plaza who are familiar with the operation of the basket and the coins it requires, payment may be relatively speedily. In fact, where barriers are not used, many regular drivers do not actually stop at the baskets but throw their coins in the basket from their slowly moving vehicle. Clearly, inexperienced users of the system can hamper the proceedings considerably (particularly if he or she does not possess the correct denominations of coinage).

Based upon the results of studies in the USA where these hopper arrangements are widespread, a single lane of a toll system may service up to 400 vehicles per hour of commuters who are familiar with the system. However, these patronage figures are exceptionally high if compared with throughput figures on most toll roads in France and Italy. Figure 3a and b illustrate the reduction in transaction and stoppage time that can be achieved by a drive through system compared to a stop and pay system.

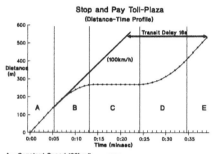

Stop and Pay Toll-Plaza
(Distance-Time Profile)

A - Constant Speed 100km/h   — .
B - Decelerate 0.3g (2.9m/ss)   D - Accelerate 0.25g (2.5m/ss)
C - Stop and Pay Manual Toll   E - Constant Speed 100km/h

Figure 3a: Distance-speed profile for stop and pay toll-collection.

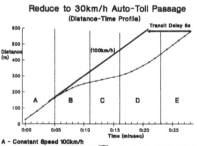

Reduce to 30km/h Auto-Toll Passage
(Distance-Time Profile)

A - Constant Speed 100km/h
B - Decelerate to 30km/h (-0.3g)   D - Accelerate at 0.25g (2.5m/ss)
C - Reduced Speed to 30km/h   E - Constant Speed 100km/h

Figure 3b: Distance-speed profile for vehicle passing through toll-site at 30km/h

Card Readers

Pre-paid journey cards (magnetic or paper-based), credit cards and more recently smart cards are now all used for toll payment purposes. All these methods of payment require that the driver inserts an appropriate card into the card reader, wait for that card to be debited (or validated) and then collect the returned card (together with a receipt, if requested) before continuing the journey.

Pre-paid cards (purchased in advance from the toll operator) are the most common cards to be used. These usually hold the "rights" for a number of journeys or the right to use the toll road at will for a particular period of time. Smart cards may also hold the same information, however they may also be used to hold '*electronic cash*', which is deducted from a card's "purse" for each toll transaction. Smart cards also have the possibility to be recharged with credit or subscription rights - although most of those in use for tolling purposes now are stored-value cards, such usage will spread now that numerous banks have adopted electronic purses held on smart cards.

The use of credit cards is not so widespread, for two good reasons:

- the value of the toll transaction is generally low and therefore credit card operators see no commercial viability in allowing credit card payments for these small values (one exception is on long-distance closed toll highway networks in Italy, Spain and France, where drivers pay a charge related to the journey distance on the network - charges of several £10's may be incurred on such journeys which make credit card payments viable); and
- the time it may take the credit card reader to validate a transaction (typically around 10-15 seconds, if dial-up lines to a card validation computer are used) make this form of payment less than attractive at toll booths, where long queues may develop if this form of payment were employed.

Based upon an ergonomic study in France, the total time required for payment using journey tickets is around 15 seconds (i.e. a processing rate of 240 vehicles per hour), whilst the total time for a credit card payment is around 22 seconds (a rate of less than 170 vehicles per hour).

Paper Stickers, Area Licences and Vignettes

A paper sticker-based system is a non-automatic means of identification, which conveys to a manual observer (or camera) visual information regarding the rights of that vehicle-user to drive on a specific road network for a specific period of time, or during certain times of day. The arrangement is illustrated schematically in Figure 4.

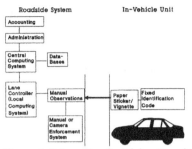

Figure 4: Schematic of a Sticker-Based System

A paper-sticker or license can only convey a small amount of fixed information and, depending on the sophistication of the sticker, the information may only be read at a short distance in slow moving or stationary traffic. This is usually achieved using brightly-coloured stickers, which are prominently displayed in the vehicle's windscreen. Even so, it is impossible to read with any degree of accuracy in fast-moving or multi-lane

The paper sticker has the advantage of being a system easy to implement and easy for drivers to understand. The difficulty with the approach lies in enforcing such a system. The licences need to be read at a distance either by a manual operator at the toll site or by a random inspection by police or another agency. It also necessary to make the permits fraud-proof and flexible enough to cater for the different subscriptions and licences that may be offered.

In Bergen, Norway, manual reading of paper stickers is used at toll sites. Special drive-though toll lanes are dedicated to those drivers possessing a paper sticker. This system is effective for enforcement but, again, requires that the road is divided into single lanes and that a manned toll booth is used. A video camera is used to take digital photographs of all vehicles that are detected as not possessing a valid licence. In Bergen, it is estimated that up to 600 vehicles per hour can be checked manually - however, it relies on the vigilance and integrity of an operator who is performing a repetitive and less than fulfilling job.

In Singapore, the Area Licensing scheme has been successfully employed for over a decade. There uses licenses of different colours to depict different access rights. The entrance roads into the central zone, where the licences apply are clearly marked by gantries which use lights to indicate when the zone is "active" (as shown in the photograph, Figure 5). Generally, these roads are two or three lane roads and there is no restriction in traffic flow. Enforcement is performed by manually by inspectors situated in booths at the side of the road. It is not known how effective they are at detecting violators across three crowded lanes of traffic. Police patrol cars are also used to check licenses through random inspections. Violators face a hefty fine and the official figures in Singapore suggest less than one percent for violations. Due to the political situation in Singapore, this violation rate does not seem unrealistic. In the early 1990's Singapore initiated a major study into the possibility of replacing the area licence by a more flexible electronic multi-lane toll collection system. Trials of three systems were undertaken over the summer of 1994 and in 1996 a decision on the replacement system was made.

Figure 5: Gantry Indicating the Boundary of the 'Restricted Zone', Singapore

In 1993, the Governments of the Benelux countries, Denmark and Germany agreed in principle to introduce a supplementary charge for heavy goods lorries using their primary roads network. It was proposed that the system would take the form of a paper licence which would be displayed in the windscreen of the vehicle. This is the basis for the so-called Euro-Vignette. The system would be enforced by automatic number plate-reading technology which would read randomly the license plates of vehicle using the road network. These would be crossed-checked with a data-base of vehicles who are known to possess the correct paper licence and violators would be fined at some later date. The key to such a system is the ability of the video licensing technology to read the licence plate of a vehicle accurately, without the need to channel traffic into single lanes. A number of systems have been tested and it is understood that the current accuracy in reading the plates is around 85%. This figure may well be good enough to enable the introduction of a system based upon the above methodology. However, relying solely on image-processing techniques for enforcement means that some deterministic ways of defrauding the system (through dirty, stolen or tampered-with license plates) may be possible - although spot-checks by operators and the police may irradicate most of the potential violators.

2.3 *Automatic road charging systems*

The different classes of automatic fee-charging systems are based upon the broad type of in-vehicle unit which is used in a particular system. This means of classifying the systems is the most useful as, generally, the functional characteristics of a system are, to a large extent, dictated by the in-vehicle equipment. Thus, the performance of the charging system under different traffic conditions; the type of payment and account options which may be implemented, where and what data is to be stored and processed are all largely prescribed by the choice of the in-vehicle equipment. On the whole, the components and the functions of the roadside system remain the same, irrespective of the system. Only where and how much data is to be processed and stored changes dramatically, depending on which class of in-vehicle equipment is employed.

Many approaches have been adopted for the automatic charging of road-use. In this section 3 distinct types of systems are described:
- systems which use short-range (5-30m) communications. These are by far the most common automatic fee collection devices and are further sub-divided into three classes (read-only tags, read and write tags and automatic debiting transponders);
- wide-area systems which use satellites or GSM (Cellular Radio) based systems as part of the charging scheme; and
- video based systems.

The technology types described here are by no means an exhaustive list of the methods available, however they are those most likely to meet the requirements for large-scale road-use pricing schemes in Europe.

"Automatic fee-collection systems" is the generic term for the procedure which allows data to pass between a device fitted to a vehicle moving at speed and a fixed roadside charging station, as the vehicle passes, for the purpose of charging a toll. It is automatic in the sense that no action is necessary either by the driver or the operator of the roadside equipment to achieve a transaction.

Each system, typically, has four distinct components:

(i)   vehicle-mounted device, referred to as an In-Vehicle Unit (IVU);
(ii)   one or more roadside interrogator stations;
(iii)   a roadside (or remote) computer system linking interrogators, for the validation, processing and storage of data; and
(iv)   an enforcement system.

Short range systems require all four of the above sub-systems. Wide area systems may not have a roadside interrogator and enforcement system at all toll collection points. Video systems generally have a combined licence plate reading and enforcement system and do not require an in-vehicle unit (except maybe a paper sticker indicating access rights for manual inspection).

To describe the different types of automatic non-stop revenue-collection system, it is necessary to consider their operational requirements. In toll-collection on interurban roads, it is generally not necessary to differentiate the charge on any basis other than the class of vehicle and (in closed toll-networks) the distance or time travelled within the network. However for road pricing the charge calculated may be based on a number of fixed and real-time parameters mentioned previously. Furthermore, for toll-collection, operators currently generally offer users a choice of payment options, which includes manual payment using cash or card machines and automatic non-stop payment using a coded tag, there is no apparent need (as yet) for more sophisticated equipment. However for more sophisticated road-use pricing, more may be required, such as anonymity of payment, various electronic payment options and centrally held accounts, timely in-vehicle, roadside and pre-trip information (to enable the driver to make the best choice for his or her travel in terms of mode, route, time and cost) and the need for reliable multi-lane operation on highly congested routes.

It should be noted that the term Automatic Vehicle Identification (AVI) is most widely used to describe almost all current systems which, in reality, rely on some form of automatic account identification (AAI). AAI is the more appropriate term, as identification of the vehicle should be irrelevant to the process, except in the case of default.

Short Range Systems
(i) Read-Only Tag

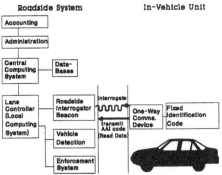

Figure 6: A Schematic of a Read-Only Tag-Based System

The read-only tag is the most common type of automatic fee-charging system and uses either post- or pre-payment AAI (Automatic Account Identification) from vehicle-users. The arrangement of such a system is illustrated schematically in Figure 6.

Read-only tags contain a fixed identification code which, when interrogated by a roadside reading device at the charging point, will convey this identity code to the roadside system. The code relates to the user's identity or the vehicle's identity or (most usually) the identity of the user's account.

The performance of the data-transfer is highly dependant upon the communications medium used to convey the information. Generally, both the range of communication and the amount of data transferred is low. All of the processing necessary for the checking, validating and performance of the transaction is carried out by the roadside system. Some of this may be performed locally (in the lane controller), such as a "blacklist" check, the

remainder is carried out (off-line) in the central system.

Read-only tags will only operate correctly if used for single lane operation. The speed of passage through the lane is dependant on what type of system is employed and on the enforcement used. At the Dartford river crossing, for example, the DARTAG is used in conjunction with a barrier and therefore allows passage speeds of only about 10 km/h; whereas, in Portugal, systems exist which allow drivers to pass through the toll lanes at up to 50km/h (although in practice many drivers' speeds may be double this) - this is enforced by video camera. In general, about 600 vehicles per hour per lane may be processed using such systems

(ii) Read and Write Tags
Read and write tags are a logical further development of the read-only tag. They have the facility to receive data from the roadside and to store this data directly on the tag or on a separate value-card (which may be interfaced with the tag whilst in the vehicle). The arrangement of such a system is illustrated, schematically, in Figure 7.

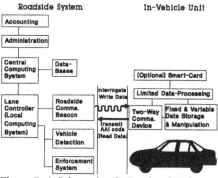

Figure 7: A Schematic of a Read and Write Tag-Based System

The ability of the IVU to receive data from the roadside system allows for the storing of information on the most recent transactions on the tag or value-card itself, thus providing the vehicle-user with an independent record of recent transactions. To date, the use of this additional function has not been widely implemented and most read and write tags operate in a similar manner to the read-only tags. However, in the road-pricing context, the facility of providing the user with an independent log of transactions may be valuable in obtaining users' acceptance of the system. It is envisaged that the user's transaction log would be read at special service points - either from the tag's memory itself or, more practically, from that on the value-cards. To date, these options have not been implemented in a commercial system. Similar road configurations and vehicle processing speed to that of read-only tags may be expected.

(iii)   Automatic Debiting Transponder

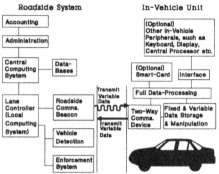

Figure 8: Schematic of an Automatic Debiting Transponder-Based System

The definition of an 'automatic debiting transponder' is one which has a relatively high level of intelligence (utilising an on-board micro-controller and associated circuitry), the capability to handle and process many kinds of data and (potentially) to be programmed to manage a number of different applications. The arrangement of this system is illustrated, schematically, in Figure 8

Such a system requires a relatively high-speed and reliable two-way data-communications link with the roadside. These systems necessarily involve more complex on-board equipment, which replaces some of the processing requirements traditionally handled by the roadside equipment. The organisations who are developing the prototype automatic debiting transponder systems have generally adopted a modular approach to the transponder's design, which will facilitate the 'adding-on' of peripheral equipment, such as smartcard readers, keyboards, displays and connections to other in-vehicle equipment. Figure

Figure 9. Transponder and Smartcard

9 is a photograph of transponder and smart card from the ADEPT project who installed systems for trial in the UK, Sweden, Portugal and Greece. The modularity of the design of the automatic debiting transponder prototypes will allow several different forms of payment with one device. The transponder offers the possibility to hold the user's credit-balance, either directly in the transponder's memory or alternatively on a separate smartcard, which is interfaced to the transponder. Other forms of payment, such as AAI (with a user-held transaction register) may also be implemented with this class of IVU.

These systems are what is perceived by many road administration across Europe as what are required for the future, where high-volume, multi-lane roads are to be tolled without having to restrict traffic. In Europe the standardisation of such systems nears completion, it may be expected that many products based upon 5.8GHz communications technology will emerge. To date only one commercial installation of such equipment exists in Europe, in Austria. This was installed in January 1995 and uses the system for both multi-lane free-flow tolling, single-lane non-stop tolling (at dedicated lanes at toll-plazas) and for 'mixed' tolling (where coin, card and transponder based tolling options co-exist in the same toll-lane). A number of installations do exist elsewhere in the world, North America and Singapore and the best examples, whilst the recently awarded multi-lane toll collection contract in Melbourne, Australia promises to be the largest such installation so far.

One interesting point of such transponders is that there is in-built flexibility, which will enable the system to be used for other purposes (such as route-guidance, parking and providing traffic information in the vehicle). Although not all drivers would wish such services, those that do could opt for these services on a subscriber basis. Moreover, the transponders offer the potential to be '*stripped down*' in functionality to operate as a simple read-only, or read-write tag - thus a family of "upwardly compatible" toll payment equipment could exist. In addition the smart card itself could also be used for other payment services and city card applications.

Thus, considering the base technology, a number of options for payment could exist depending upon the needs of the driver and the configuration of the toll collection system required. However, on roads with low daily traffic flows, the need for a full-blown system may not be apparent or economically justifiable.

Wide Area Communications-Based Systems

Wide area systems are an innovation in the toll collection arena. They make the use of two technologies which were developed for other applications, namely GPS (Global Positioning Satellites), which enable suitably equipped vehicles to calculate their location with some accuracy, and GSM (Cellular Radio), which enable two way communications over a wide area in most of Europe. These systems alone or in combination tested in the German the AGE trials in 1994/95. The systems are designed not to disrupt the flow of traffic on motorways and also to reduce the amount of roadside infrastructure required, in comparison to that needed by systems which employ short range communications. Their main disadvantage seems to be the lack of a proven and effective enforcement mechanism.

(i)   Global Positioning Systems

Global Positioning System (GPS) is a satellite-based location system, originally launched by the

Figure 10: Multi-Lane Charging Gantry from the A555 German Trials

US Department of Defence (DoD) for military applications, though it is also used increasingly for civil applications, albeit with degraded accuracy. It is now partially controlled by the US Department of Transport (USDoT) as well as the DoD. Basically, it uses the times of arrival of signals from a minimum of four orbiting satellites (out of a constellation of 24) to calculate the location of the receiver in three dimensions (i.e. including height). It is capable of accuracies of typically 100 metres. Much higher accuracies are possible using differential and phase-based techniques. However, these increase the cost of the equipment very significantly. It is important that at least three satellites are visible to the receiver and not obscured by high buildings or other obstacles, in order to maintain the position measurement.

The IVU contains a GPS receiver and must have a record of the designated locations of all charging points. For each vehicle using the appropriate highway and passing designated sites, the appropriate toll fee is deducted from an on-board account which will be most probably in the shape of a smart card. On reloading the smart card the debiting information is transmitted to the clearing agents, so that the revenues are allocated correctly.

Limited enforcement can be on a mobile, random and spot basis. In order to achieve full enforcement, some proposals foresee a short-range communications device in a mobile or road-side inspection unit, which will transmit a request payment receipt to nearby vehicles (as illustrated in Figure 11). This approach requires the use of gantries, with equipment detection and tracking of violators plus enforcement by infrared flash photography of vehicle number plates. However, this implies a full infrastructure across the highway which negates the major benefit of this approach.

Currently GPS road-charging technology is being trialled in Newcastle upon Tyne by the local university. This is used to define road charging cordons around the city. One hundred volunteers are being studied to see how they behave in both a cordon charging and a distance-based charging regime. Results of the trial will provide a good insight as to whether GPS operates well enough in urban areas to define reliable and accurate charging cordons and also on the user acceptance of such a concept.

Figure 11: Screen from the Mannesman's ROBIN system which used GPS to define virtual road-charging payment locations

Industry have become to recognise the benefit of GPS as a mechanism by which a number of charging gantries can be replaced by 'virtual' charging points. The trend seems to be to move towards the combining of GPS and intelligent transponders so that both options for charging area viable whilst the transponder communication link also offers a better enforcement option.

(ii)   Cellular Radio and Satellite Location: GSM and GPS
A road-tolling system has been tested in the German motorway tolling trials, which is similar to the GPS system described above, but using the Global System for Mobile communications (GSM) cellular radio system, as the communications link, and the GSM Subscriber Identity Module (SIM) as the tolling smartcard.

GSM is the pan-European digital cellular mobile telephone system, which has been adopted as a standard by CEPT, the organisation of European Public Telephone Operators.   It is already being

implemented in most European countries and, consequently, there is a communications infrastructure in place which (in principle) could be exploited on for tolling purposes. As described above, GPS supplies the location of the equipped vehicle and, hence, its proximity to the toll site.

On coming to a pricing cordon, the system will debit the charge from its built-in account e.g. in the smartcard.   The system can also use GSM for informing the central system, once a sufficient amount of money has been debited from the on-board account. This enables the central system to initiate the clearing process and allows the use of a wide range of payment options. The GSM system can also be used to reload an exhausted card and to update the charging tariff and locations of the 'virtual' pricing sites.   Figure 12 illustrate a GSM and GPS toll system.

(iii)   GSM Technology
Several proposals have been made for tolling systems based on charging for entering a cell.   Such systems could be based on the cell structure of the GSM communication system or alternatively use several GSM stations to improve the cell delineation.   An alternative beacon structure might be used but this implies a new communication infrastructure.   Toll charges are transferred over the communications system to the control centre.

Video-Based Systems
Automatic Number Plate Recognition (ANPR) is a variation on an automatic account identification system, which relies on the vehicle's number plate as its unique identifier.

ANPR systems are being introduced for "reading" vehicle number plates.   They process a video image, taken by a camera on a gantry, locate the number plate in the image and convert this into the appropriate alphabetic and numeric characters, without human intervention using Optical Character Recognition (OCR) techniques.

Although such technologies have been used for the enforcement of paper licence and tag/transponder based systems they have never proved effective as a fee-collection means or access control management system in their own right.   Nevertheless, if the technology for reading licence plates matured to the point of being sufficiently accurate, the system would offer a cost-effective way of implementing a central

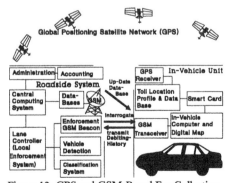

Figure 12: GPS and GSM-Based Fee Collection System

account-based or subscription-based road-pricing system. The down-side is that a record of the movements of every vehicle would be made and kept by a city or road authority - since proof that a vehicle was on a particular road at a certain time of day would be required if a driver wished to contest a particular toll charge or enforcement action.

The increase in the use of video cameras for road traffic monitoring is providing sufficient incentive for camera manufactures to move away from the strict adherence to the standards of the broadcast industry. Developments in camera technology include improvements in optical processing to provide a wider contrast range, providing clear licence-plate images even when the licence-plates are in heavy shadow or surrounded by very bright headlights in direct alignment with the camera. Figure 13 shows the new camera technology used at a toll-site in Malgara, Greece.

Developments in automatic licence-plate reading methods have been undertaken at a number of research laboratories.. Techniques based on neural network technology and the use of fast parallel-processing machines promise to produce much more robust and reliable license plate reading systems than are currently available.

3 OTHER ISSUES

The above sections have focused primarily on the mechanism by which charging can be made in a road-tolling or road-pricing scheme. There are however, other related issues that must be considered when designing a such a system.

Figure 13: Enforcement at Toll Site Location, Greece

3.1 *Enforcement*

Non compliant vehicle users must be detected and enforced. These users fall into 3 categories:
- those which do not posses the necessary in-vehicle equipment;
- those whose equipment has mal-functioned or been tampered with; and
- those users whose central account or user-held credit is in arrears.

The enforcement function requires the automatic detection (and localisation) and classification of the vehicle as well as the enforcement of non-compliant vehicles using video image recording of licence plates. Additionally for multi-lane operation the localisation of each transponder may also be necessary.

The performance and reliability of the enforcement system is critical to the successful implementation of an automatic debiting system. Both as an effective deterrent to potential defrauders of the system as well as being seen as a system which does not under any circumstance enforce drivers who correctly use the system.

3.2 *Security*

The data pertaining to the user identification (in the case of a centrally held account) or the electronic credit (user-held account) must be secure and tamper resistant. Private key techniques (such as DES) have been proposed as a means to authenticate the in-vehicle equipment and to protect data passing across the communications link. Public key encryption could be used to provide a unique '*stamp*' on the electronic credit units for use in a multi-operator/vendor system (integrated payment). This would allow each individual credit unit to be traced back to the source of purchase (Zandbergen and van Gils, 1993).

Current smart card configurations and operating systems have not been able to offer the speed of processing and data encryption needed for these applications.

3.3 *Payment Options*

An advanced automatic payment system, such as one would consider for city-wide road-use pricing should be able to offer a number of different payment options. These payment options are divided into those which require a central (roadside) held account and those which utilise an on-board (vehicle) account:

- *Pre-Payment Automatic Account Identification (AAI)*
  Centrally held user-account which is always in credit. AAI code pertaining to the account identity is conveyed from the in-vehicle tag to the roadside upon interrogation.
- *Post-Payment Automatic Account Identification (AAI)*
  Centrally held user-account which is in arrears, i.e. the payment for the road charges is made after the charges have been incurred either through direct debit or an off-line billing process.
- *Centrally held Subscription Account*
  The in-vehicle tag conveys an identity code to the roadside system when interrogated. The information relates to a subscription account held be the user which allows the user to use the road-use services for a particular period of time (eg. Travel Pass) or for a particular number of journeys (eg. Multi-Ticket).
- *Anonymous Subscription*
  The in-vehicle unit hold all the details pertaining to the subscription. This data is conveyed to the roadside system upon interrogation. In this case the user of the system may retain his or her anonymity during travel.
- *On-Board Pre-Payment (User held Account)*
  The user carries his own electronic credit with him. The in-vehicle equipment deducts an appropriate amount of this credit during the transaction with the roadside. Details of the recovered credit is passed to the roadside system, even so the transaction can be configured to be wholly anonymous.

Where information is required to be held in the vehicle for subscription purposes or where electronic credit is held by the driver, the smart card is the obvious medium for carrying this data. The smart card is a relatively secure device for carrying both subscription details and electronic credit. The read and write nature of the card enables the data or credit balance to be changed when necessary and also facilitates the possibility for a record of all transactions (user audit trail) to be held in the cards memory.

3.4 *Audit Trail*

Where toll and road pricing charges of different values are automatically collected by the roadside system it would be desirable for the user to have an independent log of the charges. This may be readily implemented in the smart card and be accessible only by the cards owner. Two alternatives exist:

- a single audit trail containing information on all the transactions for all the different applications; or
- an individual audit trail for each different application.

The implementation of an audit trail is one of the most attractive reasons for the use of smart cards in such systems. In many cases it is the only mechanism open to the user by which he or she may contest a '*wrong*' transaction or the enforcement procedure.

3.5 *Design for Occasional Users and Non-Equipped Vehicles*

In most road-use pricing systems it may only be feasible to supply electronic equipment to regular uses, those who visit a city infrequently must also be catered for in some way. The issue of non-equipped users is one of the greatest operational headaches when it comes to the roll-out of large-scale systems. Options include telephone payment of the right to use the 'priced roads' for a particular period of time, the purchase of a vignette or even the design of a low-functionality (and cheap) occasional user equipment. The issues here are not just technical but cover the general management of the system and the legal framework under which the road-pricing scheme was established.

3.6 *Legal Issues*

Legal issues associated with the enforcement of violators, the mandatory nature of the equipping of vehicles in a pricing area, anonymity and privacy and reliability/liability issues must be considered. Work is on-going in Europe to address these issues at the union level.

3.7 *Vehicle Classification*

One of the parameters which is likely to be used to calculate the due road-fee is vehicle class. This means that there must be in place some form of automatic vehicle classification scheme, which can discriminate between vehicles of predetermined classes on a multi-lane highway in real time.

Automatic vehicle classification is performed by measuring some parameters of the vehicle and comparing them to parameters stored in a database which define the classes in use. In a recent working paper for the CEC's Concertation group a table of possible parameters were compiled (Blythe, 1996), and shown in table 2. This work is now continuing in the CEC funded project ADVICE.
The technical means of discriminating the classes based upon one or more of the above classes are by no means all existing. Thus it is important to consider what is feasible before defining the pricing regime.

Generally, no clear consensus exists on the definition of appropriate parameters for vehicle classification and classification schemes. EC regulation R1108/70, for example defines 25 separate categories of vehicles but excludes motorcycles, whilst the UK national core census defines 20 categories including motorcycles and bicycles. These categories are largely based on the number of axles and wheel-base measurements. No standards have been defined for profile type vehicle classifiers nor video based classifiers. CEN TC278 now has a working group (SWG 8.1) which is developing a matrix-based approach to defining a multi-variable classification system which is appropriate to most situations.

3.8 Back- end and Central Systems

This chapter has focused mainly on the front end technology which is necessary to achieve a communications dialogue and ultimately a secure and reliable charging transaction between a vehicle mounted device and a roadside based collection system. It is worth pointing out that this is indeed not the end of the story as the central computing system and any distributed intelligence within the charging network are important elements of the chain of transaction and collection process too. It is becoming increasingly clear that these 'back-end' functions may indeed be a bottleneck to the entire process. The amount of data required to be collected for each transaction, the security and integrity of that data and the updating of white-

lists and black-lists are crucial issues which are only now beginning to be addressed for large scale implementations of toll collection and other forms of road-use charging.

What data is stored that could be traced back to a driver/user and who has access to this information are important questions that operators must address. In the past such issues were not so relevant as users 'opted in' to electronic payment systems by choice as a replacement for paying by cash, however now the option to choose may well be limited as in many cases ALL regular users of the road network will be mandated to 'opt in' to the system. Here forms of anonymous electronic payment using smart cards and/or electronic journey tokens are important as a means of allaying such concerns and providing the users with a choice (even if 95% of the users do not care about such things a vocal 5% could be the difference between success and failure).

Apart from the transactional data and user lists which must be passed backwards and forwards from roadside to central systems there is also another potentially difficult to handle data stream, that of enforcement images. In a large scheme where there are potentially millions of transactions per day, the recording of just one potential violators image could be up to 1Mbyte of data (converted from video or digital photograph), if only loss-less compression of images is allowed (i.e. the data is perceived not to have been tampered with in any way - from a legal viewpoint of evidence) then a large data transfer and storage capability is required which imposes fairly stringent requirements on the system. If only a few percent of images were stored as potential violators this could run into tens of thousands of images per day - these images may need to be stored for a relatively long period as the follow-up investigation and enforcement procedures could take weeks and if it goes through the legal system, months - here clearly the industry can learn from others who have implemented cameras for other forms of traffic regulations enforcement.

3.9 Interoperability

Interoperability of road-use pricing systems is most likely dependant upon the outcome of the CARDME concerted action which is addressing this issue from an pan-European toll collection perspective. In Europe such operation is desirable.

Indeed similar considerations can be seen being addressed in North America, where different toll-road authorities share a toll tag and revenue collection and apportionment systems. In Europe the MOVE-IT project has been successful in defining some degree of contractual interoperability between operators for tolling, whilst the ADVICE project is aiming to establish common practices and procedures for classification and enforcement of road pricing and tolling systems.

3.10 *Standardisation*

Within Europe there has been a concerted effort to establish standards for the new generation of advanced tolling and road-use pricing systems. The process was stalled for a number of years due to the reluctance of some suppliers and National administrations to accept that standards were needed (rather than nationally preferred and supported systems), however the sterling efforts of the Commission, CEN and validation projects such as VASCO have overcome this understandable reluctance. In 1997 some of the key elements of the charging equipment were finally approved by CEN TC278, namely the short range link (physical, logical and application layers) based around a 5.8GHz communications. The Storebælt toll bridge will be one of the first recipients of equipment which meets these new European Standards Other standards will follow in mid 1998. The agreement of these short-term (ENv) standards will hopefully open up the market for such equipment as now a number of suppliers can now supply equipment to the same specification - this should also reduce the cost of the equipment (due to competition) and also guarantee second sourcing. Indeed the GTS (global toll specification) that has been established by a number of European suppliers (Boche, CGA and Combitech) to demonstrate how interoperability can be facilitated using multi-sourced equipment. What is still missing is a comprehensive set of agreed application data elements and procedures to define standard applications for both the tolling/pricing functions and for the use of the associated smart card - which would open the way for systems that support a number of value added services.

4. SPECIFIC TOLLING AND ROAD PRICING CONFIGURATIONS

4.1 *Road-Use Pricing*

The pricing of road-use as a means of dynamically charging drivers for the use they make of the roads has been one of the most widely discussed options for more that three decades now (Smeed, 1964). However, to date, few schemes have been tried in Europe, and even fewer deployed on a large scale.

The basic problem in most cities is that to the car user, the road-space is a 'free good'. The roads can be used without having to pay a charge that is directly related that the extent or time of use. The, so-called 'buffet syndrome' prevails, i.e., having paid for up-front the costs for a car the user has little incentive to restrain usage but rather gorge oneself.

Clearly the political, social and economic consequences of charging driver's for something that until now they have perceived as free at the point of use is difficult, although it is noticeable that the opposition to tolling on the UK motorways is no longer overwhelming. Nevertheless only a few examples of road-use pricing in Europe can be cited. The most basic form of road pricing, and indeed the one most commonly considered is that of simple cordon pricing, whereby a set of changing points is established which form a water-tight boundary at which drivers must pay to enter the cordon. In most cases due to traffic volumes it is not feasible to introduce conventional, stop and pay toll plazas and thus some automatic form of payment is employed which does not require the driver to stop and pay. In Trondheim, Norway an automatic toll ring has

Figure 14: AVI and Manual Lane Toll Plaza at Randheim, Trondheim

been in operation since 1991 (Hoven, 1996). 60,000 vehicles are equipped with AVI (Automatic Vehicle Identification) tags which enable roadside antenna to read the identity of the vehicle from the tag, figure 14.

Money is then deducted for the passage from a pre-paid account the driver holds with the toll-authority or by post-payment (through direct debit from the drivers bank account). Although not a particularly sophisticated system it is effective. In a study carried out by the CONCERT project, the scheme has reduced traffic passing into the cordoned area by 8-9% in the peak travel periods, public transport patronage has accordingly risen by 7%. Similar schemes using manual and mixed manual/AVI payment also exist in Bergen and Oslo respectively.

A number of other cities have been attracted by the Trondheim approach. Gothenburg and Stockholm have both considered toll-rings as a means of road-use pricing (in Gothenburg a wide area system using GSM or GPS is being considered, in Stockholm the plan was to use an intelligent transponder based system however this project, which was quite advanced, has recently been cancelled) as are other historic cities who's plans are not yet as advanced.

So far, the most comprehensive example of advanced electronic road pricing may be found in Singapore where 3 systems were trialled in 1994. Finally the Philip Consortium was selected to build and electronic toll ring to replace the current area licensing scheme (figures 15a and 15b).

A more sophisticated form of road-use pricing was tested and evaluated in Cambridge under the ADEPT project. Here the cordon was again defined by microwave beacons, but instead of charging the driver as he or she passed the cordon, the in-vehicle equipment measured the average speed of the vehicle and deducted units from the drivers smart card only if the vehicle was in and contributing to congestion (Clark, Blythe, et al, 1995). The trail was only a technical and functional demonstrator which showed that sophisticated forms of road-use pricing were technically feasible Figure 16a shows the generalised charging algorithm used and figure 16b one of the roadside entry beacons sighted around Cambridge.

Figure 15a Philips Consortium Gantry from Singapore Trials

Figure 15b Waterproof Housing for Transponder mounted on Motorbike from GEC-SEEL Trial in Singapore.

Trials also recently started in Leicester on road-pricing. 100 triallist are using transponders to enter the city centre where they pay a charge based upon the time of day and level of congestion. Early indications suggest that the scheme would be a success if implemented on a large scale, however the charges required to affect a drivers behaviour would be quite significant

4.2 Access Control

Access control systems have also been deployed as a means of managing demand. The most prominent schemes in Europe are those in Barcelona and Access control and parking management measures have also been trailed as a way of controlling demand. In the GAUDI trial in Barcelona (Hayes, 1994) a smart card based access system was used to restrict access to certain residential areas of the city. The cards lowered hydraulically activated barriers in the road. Such a scheme received high levels of acceptance from the users.

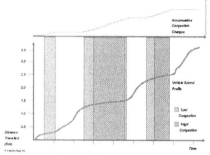

Figure 16a: Congestion charging profile as used in the Cambridge Congestion Metering Trial

Figure 17a: Bridge-Mounted Toll Collection Equipment

Figure 16b: ADEPT 'Entry Point Beacon' from Cambridge Congestion pricing Trial

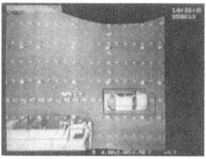

Figure 17b: Output of Vehicle Detection and Classification System

Access control has the advantage of not necessarily requiring payment at the point of use. The rules for access may be based upon need, or whether you live or work in the controlled area or not. In many cases the technology required to operate such a system may be less than that required for road pricing. Stickers and read-only tags are the favoured methods.

4.3 Road Tolling

The charging of tolls on motorways has emerged on the political agendas of many countries. In most cases the move is to toll motorway networks that have until now been free to use. Such a policy obviously raises some difficult political problems,

but also some practical ones. Telematics can only be used to solve the practical ones and these relate to the need to collect tolls on multi-lane highways, without requiring the driver stay in lane, slow down or stop and without the possibility of introducing toll-plazas and manual means of payment for the drivers. Such systems, known as multi-lane are technically complicated but are now close to achieving the level of performance necessary for large-scale implementation, (Blythe and Burden, 1995).

Many countries in Europe, most notably Germany, Austria, Italy and the UK have embarked on ambitious schemes to test and eventually install systems elsewhere there is great activity in Australia, North America, Japan and S.E. Asia. So far Austria is the only country to actually introduce such a system which also incorporates a novel and effective on-line video classification system (figure 17a and b).

Germany trailed 10 possible systems in 1994/95 the results of which are now in the public domain (Stappart, 1996). The UK is also currently in the

Table 2: Parameters on which class of vehicle could be based

Vehicle Type	Other	Vehicle Dimensions
• motorbike • small car • large car/small van • large van/mini-bus • small/medium/large HGV • with/without trailer • number of axles • various bus sizes	• Engine capacity • Environmentally friendly/low emissions • High Vehicle Passenger Occupancy (HOV) • Type of goods carried • etc.	• length/width • height • length (including trailer) • number of axles • etc.
Weight • unloaded weight of tractor unit • total weight of vehicle • weight of trailer unit • axle loading • etc.	**Type of journey** • commuter • professional driver • heavy goods • exempt • occasional driver	

final throws of evaluating two systems. Apart from the obvious objectives of the UK trials, in terms of raising revenue and putting the operation of the roads in the hands of private consortium, the toll system is to be used to manage traffic demand. The proposed configuration of the system so that a small charge is made at every junction on the network. Such a distance-based charge may well act as an antidote to induced traffic, although only a real trail would validate this hypothesis

5. SUMMARY

Technology for road -use pricing and tolling using automatic means has been continually under development for the past 15 years. Recently the trend has been to look at integrated systems which are flexible and may be able to support other traffic management and information systems, this is emphasised in the examples of systems given in this chapter. Now the main emphasis is on obtaining standards and common ways of working to achieve integrated, interoperable systems. This is where the new challenge lies and where the opening up of the electronic toll market will ultimately succeed or fail.

6. REFERENCES

ADEPT V2026 1996. Final Report, *ADEPT Consortium, Commission of the European Communities, DG-XIII (C-6),*

Blythe, P.T. and Burden, M.J.J 1994. Electronic Toll and Traffic Management - New Developments in Technologies and Systems. *Proc. Asia Roads and Highways, Hong Kong,* September

Blythe P.T 1995. The Use of Smart Cards in Road Pricing and Tolling Systems. *Proc. IEE Colloquium on Electronics Techniques for Road Pricing and Tolling,* London, March

CEC 1994. *'The Transport Telematics Programme' .Commission of the European Communities, DG-XIII (C-6),* Brussels, Belgium

Blythe, P.T. and Hayes, S. 1995. Area 1: Demand Management. *Chairman's report. .Commission of the European Communities, DG-XIII (C-6), Brussels,* Belgium

Hayes et al., 1994 ATT Zone Access Control Implementation in Barcelona & Bologna, *World Transport Telematics Congress,* Paris, Nov.

Hoven, T 1996. Experience with the Norway Toll Ring Project. *Proc IBC Conference on Electronic Payment Systems in Transport,* Amsterdam, March

Stappart, K-H, 1996. Field trials on electronic fee collection on Motorways in Germany. *Proc IBC Conference on Electronic Payment Systems in Transport,* Amsterdam, March

Thorpe, N.T., Clark, D.J., and Smith, F. 1997. Experimental Design for an investigation of drivers' behaviour in response to a GPS-based automatic debiting system. *Proc. UTSG Conference,* Bournemouth, UK, January

Operation and Maintenance of Large Infrastructure Projects, Vincentsen & Jensen (eds)
© *1998 Taylor & Francis, ISBN 90 5410 963 7*

Management, procedures, risks and exception handling

A. Skadsheim
Avenir A/S, Oslo, Norway

This paper highlights challenges and solutions related to the design and operation of toll collection systems in the 4 following areas:

-Management: This part of the paper reflects some important challenges in establishing and operating a large toll collection system. When operational, well planned service, maintenance and customer relations are among the most important tasks.

-Procedures: It is necessary to develop operational procedures as a part of the design. Well planned procedures prevent customer dissatisfaction, assure a smooth transition from the planning phase to the operational phase and prepares for the special tasks and problems during the commissioning of the system.

-Risks: Risks related to project management, technology, operation and external factors are discussed.

-Exception handling: As toll collection systems become more and more automated, less people are available to assist and more technology may not function. This part of the paper will describe some of the most common problems and present some proposals on how these exceptions may best be handled.

1. INTRODUCTION

These are extensive topics and only some chosen problem areas will be highlighted. The presentation is meant to of general character, but with examples from real installations including of course the Storebælt bridge.

2. MANAGEMENT

The management of a toll collection system is not dramatically different from any other business. The principles according to which the operation shall be run is mostly decided during the project phase in advance of the opening of the system. We will therefore highlight some of the most important factors that decides the operational management.

2.1 Project management

2.1.1 Legal and institutional issues

There are many institutions which will influence the design and operation of a toll system. In figure 1 are listed examples of institutions frequently involved in a typical toll project.

Institution	Topic
Ministry of transport	Principles, toll fee structure, enforcement legislation and cross border legislation
National Road Administration	Depending on the country the involvement in toll projects are from nothing to everything
	Signs
Police	License plate register/enforcement
	Traffic safety
Consumer organisations	Handling user interests
Data "inspectorate"	Regulation of computer registers and storage of video recordings
Standardisation bodies	Technical standards, issuer identification codes...
Labour inspection	Working conditions
Telecommunication authorities	Concessions for use and type approval for radio systems

Figure 1.

To achieve a customer friendly, efficient and safe operation, it is also necessary to establish contractual and physical interfaces with a series of private and public companies and institutions

2.1.2 Operational interfaces to external bodies

Clearing houses	Validation of payment cards
	Clearing of turn-over between service providers
Other card issuers	Oil companies' cards are normally handled separately from credit cards
Banks	Payment from cards and subscriptions
	Counting of money from plaza
Large customers	Invoicing and other information via EDI network
License plate register	Violators are identified by license plate
Other toll roads, ferry companies etc.	Common subscriptions
	Combination tickets
	Direct clearing

Figure 2.

2.1.3 Conflicting interests

During the design of a toll system, compromises have to be made between conflicting interests. This may not be a problem if the consequences of the decisions are clear and accepted. Examples of such conflicts are:

- Specially designed software is often risky and costly, but will have to be added to standard solutions to cover special needs
- Toll stations should be compact and unmanned, but sophisticated technology which may be expensive and lacks long term proof of operation, is needed to accomplish this
- The service level should be maintained at the same time as personnel is reduced to reduce cost.
- Considerable savings in investments may be made by splitting the supply into several contracts. This does however very often result in both technical and contractual interface problems. In most cases it is therefore worth the extra initial investment to have only one contract partner.

2.2 Operational management

A special problem related to some toll projects (and other types of projects as well) is the change from a planning organisation to an operational organisation. Key personnel may leave if they are not at an early stage guaranteed an attractive position in the new operational organisation. Also many large toll stations and their operational offices on inter-urban road projects are located locally, at a different location than during planning phase.

The number of employees on a toll station or a tolling company varies much from project to project. The level of automation varies very much not only depending on the level of automatic payment systems, but also how administrative tasks are being handled. As the need for personnel is greater during the start-up phase than later, it may be expensive to employ all the people necessary to operate the system during start-up. Sub-contracting of tasks should therefore be considered. Tasks that may be sub-contracted are for example:

- Facility management
- Manning of toll booths
- Service and maintenance
- Transport and counting of money

If manning of booths, service and maintenance is sub-contracted, it is very important to administer these service activities in a way to optimise the relation between available support personnel and the need for support. By reducing the service response-times outside normal office hours the price may be considerably reduced. An important clause in the contracts is therefore to be able to re-negotiate or re-tender the contracts at regular intervals, based on experience.

The primary types of personnel to operate the system are toll attendants and customer relation personnel. The rest of the organisation should be considered as support personnel. A motivated and well trained toll booth operator will have a significantly higher capacity and fewer complaints than the opposite. The number of people needed to man the toll booth also varies from country to country depending on the local labour condition regulations

An important observation is the increased need for customer relation personnel in toll systems with large subscription schemes and even more if video enforcement systems are introduced in non-barrier systems. The work relating to answering questions, changing subscriptions and handling complaints must not be underestimated. It can however be made much more efficient if the technical solutions are based on needs defined as a part of the system requirements. This includes automatic and quick access to subscriber registers etc. from all customer support work stations. Correspondence between toll company and customers should be automatically generated/ scanned and available from the individual subscriber register. This will enable many cases to be handled and closed while the customer is on the phone.

3. PROCEDURES

The technical installation must be well harmonised to the operational organisation and vice versa. Tuning the technical solution and the organisation must be a continuos process both during the project phase as well as during operation. The procedures must include at least the following main items:

3.1 User procedures

Describes the operation of the toll station as seen from the customer point of view; The signs on the road leading to the toll station, the signs in the toll station and the meaning of these; The use of dynamic displays; Alternative types of tickets, subscriptions and methods of payment. The user procedures also describes the functionality of the equipment with which the user interact and what he should do in case of problems.

These procedures are used as a basis both for marketing purposes, user instructions as well as for test procedures.

3.2 Operational procedures

Describes the installed equipment and its functionality with regards to operation, exception handling and simple error detection and correction.

These procedures will be used both as a basis for job descriptions for operational personnel as well as test procedures.

3.3 Exception handling and dispute management

A situation which deviates from the normal operation shall be handled as an exception, and all exceptions shall be handled according to predetermined procedures. If there is a dispute between a user either when he is in the lane or if he contacts the toll company by phone or mail, standard procedures shall be available.

Exceptions and disputes are becoming more and more common with the introduction of automatic systems like subscription systems and video enforcement systems. This means that the traditional contractors of toll equipment do not necessarily have the know-how and/or systems available to handle such situations. It is therefore very important to define these areas in sufficient detail at an early stage to be able to include this functionality in the tender specifications.

These procedures will be used both as a basis for test procedures and as job descriptions for involved personnel.

3.4 Daily operation

Describes for example the following tasks:

- Disposition of lanes
- System monitoring
- Error handling
- Shift administration and "cashing-in" routines
- Money management
- Preventive maintenance
- First line service
- Refill consumables
- Card and tag administration
- System administration

3.5 Marketing, sales and customer contact

The primary marketing function include marketing and information related to the road project (bridge, tunnel, highway) in general and the toll collection syhstem in particular as well as focusing on the most favourable methods of payment.

In addition to sales in the toll booths the sales function often include sales of subscriptions either directly or via distributors and a local sales office. The sales function also include automatic renewal of subscriptions as well as reimbursement of multi-trip cards and prepaid subscriptions.

3.6 Safety and security

A significant number of procedures must be produced to handle the security and safety related parts of the operation. These procedures include for example:

- Traffic safety
- Employee safety
- Entrance control
- Robbery protection
- Protection against vandalism
- Data security
  - Computers
  - Cards and tags
  - Network and communication
- Emergency situations

3.7 Administration

The administrative procedures are more or less the same as in any organisation and will not be mentioned her.

3.8 Start-up procedures

The time immediately before and after the opening of a new toll system is often very different from a normal operation. Some points to note are the following:

- Marketing plan
- Special marketing to attract sufficient number of users already from the beginning
- Special marketing to have a sufficient number of subscribers to avoid capacity problems in the manual lanes
- Design and produce all forms necessary long before the opening (subscriber forms, invoices, many types of receipts, ticket formats and layout etc.
- Design of cards, tags, signs, toll booth , etc.
- Initialisation and distribution of tags/transponders and cards
- Extra personnel
- Employment and education

4. RISKS

There will always be elements of risk in large toll projects. New technology may give operational savings and improved customer service while price competition may favour a less experienced supplier and so on. Risks can not be completely avoided, but by identifying the risk factors and monitor these and build the necessary safety nets, problems may be detected early and consequences can be minimised.

4.1 Equipment

There are two major areas of risks regarding equipment.

4.1.1 Introducing new "high tech" equipment

To accomplish objectives with regards to efficiency it is often necessary to include automatic systems which may not be completely tested in the right environment. If such equipment fail, it will both mean loss of confidence among the users as well as potential loss of income and/or added costs.

For Storebælt we consider both the classification system (based on stereoscopic view) and the new version of the tag system to be among such potential risk factors. These systems will therefore be subject to extensive testing and back-up solutions are prepared.

4.1.2 Special design software

This is very familiar problem in many projects. This risk must be reduced by trying to reduce the need for special software to a minimum at the same time as the ability to handle special design software should be a deciding factor when choosing a supplier, if such software is considered extensive.

4.2 Operation

During operation the system and the organisation may be vulnerable to situations that may cause capacity problems, added costs, customer dissatisfaction etc. Some keywords may be:

- User problems (by drivers or operators) due to weaknesses in functional and physical design
- Sabotage or fraud
- Capacity problems due to traffic volume or lower capacity than planned for
- Loss of data due to errors in equipment, network, software or operational procedures

Many (hopefully most) of these problems may be avoided by preparing operational procedures and perform adequate testing before commissioning.

4.3 External conditions

Legal and institutional issues mentioned in the beginning of this paper should be handled early in the project and should not be considered as risks, merely conditions. If there exist other toll system in the country, many problems have already been handled. Introducing new technology may however require new considerations by relevant authorities. This is often the case when introducing tag-systems, video enforcement etc.

Other external risk factors, include technical problems related to physical interfaces to external bodies or other subsystems. It is therefore important to include in the contract with the system supplier that developing/testing such interfaces is a part of the supply.

4.4 Project management

4.4.1 Delays

There is always possibilities for delays in the production and implementation of a toll system. A supplier is normally more positive to short delivery times in the tender period than later in the project. The progress of the project should therefore be based on payments related to several clearly defined milestones throughout the project.

4.4.2 Added costs

When producing a tender specification it is important to focus on functionality to allow for technical solutions proposed by the individual supplier. Such an "open" specification requires a very clear and complete description to avoid the contractor interpreting the requirements different from what was the intention.

By using adequate budget risk management procedures it is possible to identify those elements of the project that have the biggest finical risk and put extra effort into controlling these cost elements.

4.4.3 Lacking performance

Lacking performance should not come as a surprise on the opening day. Extensive tests simulating real loads must be prepared and made to avoid major problems during start-up and operation. Tests should be made both at the factory and on -site, module by module as well as integrated tests performing real transactions on the system.

In addition to the normal tests performed after production and after installation, an "In Service Test" is recommended after the commercial opening of the system. Such a test may for example be made during a 100 days period in which no major errors shall occur. If any such errors occur, the 100 day period is restarted. Such a test is a major incentive to the supplier if it is connected to a payment schedule.

4.4.4 Unidentified needs

As a project progresses additional requirements may be identified that are not included in the contract with the contractor. Usually these needs are identified when the operational procedures are detailed. It is therefore important to start this work at a very early stage in the project, as added functionality becomes very much more expensive after a contract is signed compared to if it was included from the beginning.

5. EXCEPTION HANDLING

Traditionally toll stations have been manual where most problems have been solved by toll attendants assisting the driver in the lane. As these installations become more automatic, less people are available to assist and more technology may not function. When producing the operational procedures, exception handling will be a major part.

Most exceptions may be put into one out of four categories:

- violations
- disagreements
- misunderstandings and customer errors
- system errors

The types of exceptions will to a certain degree depend on payment methods, types of tickets and technology. It should in particular be noted that there are significant differences in the exception handling in systems with and without barriers, and urban systems versus inter-urban systems. The different exceptions can be identified through experiences from other toll systems and through a "what if.....? - analysis".

In barrier operated systems exceptions may cause long delays if it occurs during rush hours. Depending on the traffic volume it may be decided either to solve the problem while the vehicle remains in the lane or let the vehicle through and solve it later. To be able to let a vehicle through without paying, necessary documentation of passage is needed. This situation resembles situations where there are no barriers and illegal passages are documented by video.

Even if there are a hundred different situations or types of exceptions, the problem may be reduced to a few methods of handling:

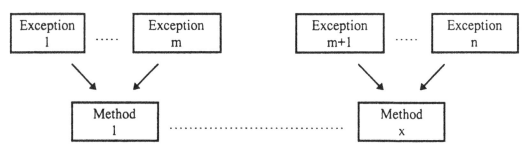

Figure 3. Simplification of exception handling.

If the problem can not be solved in the lane, the vehicle must either be led to the roadside after leaving the toll lane, he may be directed to a local customer office or he may pay later if this is agreed with the operator.

An alternative solution to letting vehicles through

before valid payment is received, is to introduce a new toll booth at the exit of the merging area. This booth could both be used as a normal manual lane, but also as a booth where exception handling may take place. The drawing below show the principle of such a solution:

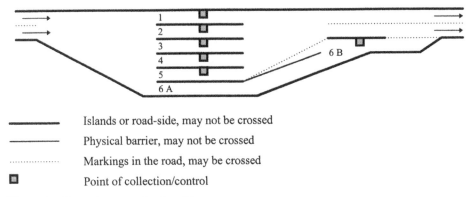

| | Islands or road-side, may not be crossed |
| Physical barrier, may not be crossed |
| Markings in the road, may be crossed |
| Point of collection/control |

Figure 4. Alternative exception handling

A vehicle entering lane 6 is lead directly to 6 B
where a normal manual payment may take place.
Any vehicle having a problem in automatic lanes 1-5
may be lead to lane 6 B for a manual handling of the
problem.

6. CONCLUSIONS

The main conclusions from this paper is the
following:

1. The design and implementation of a toll
   collection system must be a parallel process
   between functional and organisational
   development
2. The operational procedures should be developed
   early enough to include the technical solutions
   needed for these procedures already in the
   requirements specifications
3. The management tools in a toll collection system
   are to a great extent decided during the design
   process, but flexibility can be achieved through
   sub-contracting labour with short re-negotiation
   intervals.

Operation and maintenance management

Operation and Maintenance of Large Infrastructure Projects, Vincentsen & Jensen (eds)
© 1998 Taylor & Francis, ISBN 90 5410 963 7

Optimisation of management of operation and maintenance of structures and installations

Jens Sandager Jensen & Jens Kristian Ørnskov
COWI Consulting Engineers and Planners AS, Lyngby, Denmark

ABSTRACT: Optimisation of O&M of structures and installations in traffic corridors is a complex multidisciplinary effort, which is not only limited to a traditional economic cost benefit analysis. Environmental, customer satisfaction and image considerations shall also be integrated.

It is not possible to outline a single parameter of highest importance. However, the utilisation of a well structured O&M management model with a clear definition of the various management tasks at strategic, tactical and operational levels is a first major step in an optimisation process.

If the latter is ensured in combination with the use of:
*   efficient management methods and tools
*   continuous O&M development of knowledge and competence
*   the latest O&M assessment and rehabilitation approaches
*   the building up of an O&M organisation based on a well structured management model
*   utilisation of the international capabilities of O&M Management Consultants, Consulting Engineers and Contractors as well as local skills
optimisation of the O&M of structures and installations can be achieved.

1. INTRODUCTION

Management of operation and maintenance (O&M) of traffic corridors is a complex multidiscipline effort. Irrespective of whether the organisation is a large in-house infrastructure owner and/or operator organisation or whether the organisation is a narrow infrastructure owner and/or operator management organisation, which only undertakes management and contracting activities, utilisation of a structured management model is required.

A management model that is based on subdivision of management tasks into strategic, tactical and operational matters, will allow establishment of a proper decision making basis before decisions are taken. Figure 1 defines the basic management model for a combined O&M organisation.

At the strategic level the following matters are dealt with:
*   definition of customer service level

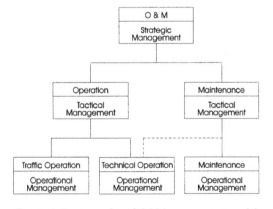

Figure 1. Basic combined O&M management model.

*   development of strategies
*   decision on level of in-house and external work
*   image management.

At the tactical level the following matters are dealt with:
- establishment of targets and evaluation of results
- development of policies
- development of O&M programmes
- operation of management systems
- organisational development.

At the operational level the following matters are dealt with:
- detailed planning of O&M operational activities
- implementation and progress reporting of O&M operational activities
- documentation and quality assurance.

The basic combined O&M model allows optimisation of management to take place at all levels. Hereunder adjustment and/or revision of strategies and policies based on an evaluation of results. Utilisation of Operation Management Systems (OMS) and Maintenance Management Systems (MMS) allows a complex and total optimisation analysis to be carried out at both strategic level and tactical level.

As part of the optimisation analysis a sensitivity analysis can be included for different alternative strategies and policies. Hereunder the influence of adjustment of one or more traffic corridor key parameters on the other key parameters, where some of the key parameters are reliability, availability, traffic disruption, cost, maintainability, environment, customer service level and image. Other important basic parameters are the length of the optimisation period and the discount rate/real interest. Figure 2 illustrates the influence of a the principal traffic disruption strategy "no traffic disruption" on other key parameters, assuming a short time horizon. Figure 3 illustrates the similar influence for a long time horizon.

2. O&M DECISION MAKING

The O&M decision making process in connection with each traffic corridor is unique. In order to structure the process the management has to define, which matters are dealt with at which level in the O&M organisation. Figure 4 illustrates how activities can be categorised in connection with a combined roadway and railway traffic corridor between two islands (fixed link) with a combined infrastructure owner and operator organisation.

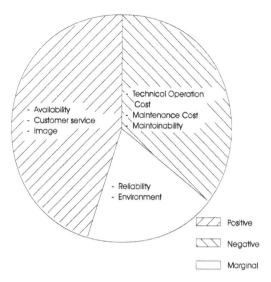

Figure 2. Strategy: No traffic disruption. Short time horizon. Influence on key parameters.

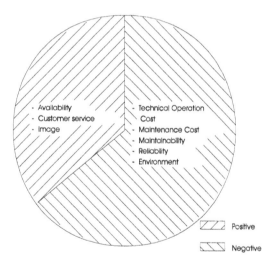

Figure 3: Strategy: No traffic disruption. Long time horizon. Influence on key parameters.

The categorisation has been defined based on the following criteria:
- Category 1 activities are characterised by a minute to minute customer satisfaction relationship
- Category 2 activities are characterised by a day to day customer satisfaction relationship
- Category 3 activities are characterised by a

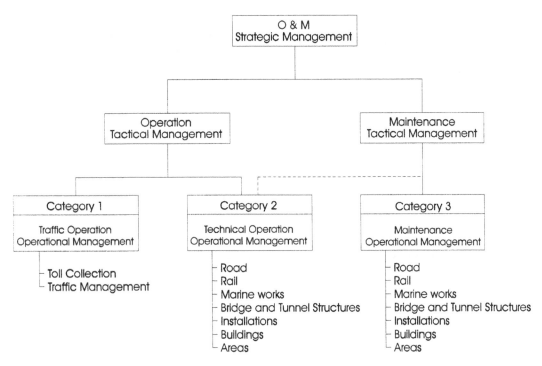

Figure 4 Categorisation of O&M activities for a fixed link with a combined roadway and railway infrastructure owner and operator organisation

minimum customer relationship when optimised management of O&M is performed.

2.1 *Strategic Level*

At the strategic level matters are dealt with, that require input to the decision making basis from all three categories of activities:

- Traffic operation
- Technical operation
- Maintenance.

At the strategic level the following key matters are typically dealt with:
- definition of customer service level
- development of operational strategies
- development of maintenance strategies
- development of rehabilitation and extension strategies
- decision on level of in-house and external work
- image management
- definition of environmental strategies.

2.2 *Tactical Level*

O&M activities are at this level carried out by two independent sub-organisations:
1. An operator organisation with the primary responsibility to ensure customer satisfaction, traffic safety, etc. and thereby to generate income. Responsible for both traffic operation and technical operation within the framework defined at strategic level.
2. An infrastructure owner organisation with the primary responsibility to ensure reliability, availability and maintainability within the framework defined at strategic level.

At the tactical level the following key matters are typically dealt with:
- establishment of goals and benchmarks
- development of policies
- development of O&M programmes
- evaluation of results
- implementation, upgrading and operation of management systems
- organisational development in the form of training and courses.

2.3 *Operational Level*

At the operational level the following three separate categories of activities are performed:
1. Traffic operation in the form of traffic management and toll collection.
2. Technical operation of road, rail and other infrastructure.
3. Maintenance of road, rail and other infrastructure.

During the planning process at both the tactical level and the operational level the relationship between O&M programmes and activities in connection with technical operation and maintenance has to be considered.

3. MANAGEMENT STRATEGIES, METHODS & TOOLS FOR STRUCTURES AND INSTALLATIONS

3.1 *Operation strategies*

3.1.1 Traffic operation strategy

Operation strategies vary significantly from operator to operator. The type of traffic, i.e. road, rail or combinations thereof, as well as geographical functional traffic characteristics (urban, regional & international) very much influence the choice of strategy.

In case of a combined operator and infrastructure owner organisation, the strategy will typically differ from the strategy defined by an infrastructure owner who is selling tickets to road traffic based on supply and demand considerations and rail km to different competing train operators.

This paper will not elaborate further on this important subject.

3.1.2 Technical operation strategy

Technical operation strategies are closely connected to both traffic strategies and maintenance strategies in relation to reliability, availability, safety, service level, image profile, etc.

This paper will not elaborate further on this important subject.

3.2 *Maintenance strategies*

3.2.1 Corrective Maintenance Strategy

Corrective maintenance is carried out in the form of repairs when functional requirements of structures and installations are no longer met. This maintenance strategy will generally result in low cost maintenance of structures and installations during the early part of their service life and high cost during the subsequent later part of their service life.

There is a high risk of associated indirect traffic corridor disruption cost and an uncontrollable accumulation of a backlog of maintenance work.

3.2.2 Periodic Maintenance Strategy

Periodic maintenance of structures and installations is performed, inclusive of preventive maintenance, as a way of reducing the frequency of failure to meet functional requirements. This maintenance strategy will generally result in a risk of cost penalty as excess maintenance work may be performed. However, there is a significant reduction in the risk of indirect traffic corridor disruption cost and accumulation of a backlog of maintenance work compared with the corrective maintenance strategy

3.2.3 Condition Based Maintenance Strategy

Condition based maintenance of structures and installations is aiming at being able to predict when failure can be expected. This maintenance strategy will in principle allow maintenance work to be performed before failure occurs, based on inspection and monitoring programmes and experimental/numerical deterioration modelling, hereunder the associated consequences on reliability with time. As a consequence, the risk of associated indirect traffic corridor disruption cost and accumulation of a backlog of maintenance work is minimised.

In case of important structures and installations subjected to very aggressive environment and/or extreme functional requirements an investment in monitoring systems will in many cases be justified.

3.3 *Management Systems*

The final goal is to develop integrated computerised operation and maintenance management systems. This will in principle allow the O&M strategic management to establish the optimum decision basis.

3.3.1 Operation Management System (OMS)

Within the railway sector traffic management systems, inclusive of train schedule planning, have been used for many years. These systems are continuously being optimised.

This development is also occurring within the roadway sector in order to assist increasing the road capacity and to improve the safety and environmental aspects.

This paper will not deal further with these important aspects.

3.3.2 Maintenance Management System (MMS)

Optimum maintenance management methods very much rely on the availability of knowledge on the performance of structures and installations, and hereunder also the causes of failure of identified failure modes.

Very efficient monitoring and inspection techniques have during recent years been developed which allow high quality monitoring to take place at relatively low cost. As a consequence technical and economic optimisation in relation to a optimum condition based maintenance strategy can be carried out on both new and existing concrete and steel structures.

The development of condition monitoring techniques, function testing, visual inspection, etc., which allow reliability centred maintenance of equipment and systems, with relatively short operational life compared to structures, also contributes significantly to the maintenance optimisation

Condition based bridge and tunnel maintenance is considered to be a well proven 15 - 20 year old method. The concepts of the present generation of maintenance management systems are being reviewed in many countries in order to include the risk element more directly into the planning process and improving the diagnosis part, the cost part and the associated programming.

It cannot be stated generally what development strategy in relation to maintenance management systems is the optimum. In case of an old infrastructure, which no longer has a uniform bearing capacity, maybe because of the occurrence of natural disasters such as earthquakes and flooding, and where a general increase in traffic intensity, weight and velocity has resulted in much higher traffic loads today compared with the original design traffic loads, it is beneficial to concentrate the MMS development on integration of structural reliability and safety aspects.

In case of relatively new structures in industrialised countries which have been designed to relatively new design traffic loads resulting in sufficient bearing capacity in relation to the existing traffic pattern and the expected traffic patterns within the next 10 years, it is more beneficial to concentrate the MMS development towards being able to carry out a total optimisation analysis.

In the latter case the optimisation process basically has the goal to ensure that optimum preservation of the investment takes place, with due consideration to other key parameters such as availability, traffic disruption, maintainability, environment, customer service level and image, and in this context providing the fund raisers with the results of a proper optimisation analysis including life cycle analysis.

4. O&M KNOWLEDGE AND COMPETENCE DEVELOPMENT

Optimisation of O&M management requires, apart from the use of a O&M management model and very efficient management methods and tools, the use of the latest technical knowledge within maintenance and repair of structures and installations as well as a continuous development of the organisation. It is a must to transfer technical knowledge into competence and motivation of the individual employee and groups of employees by setting up training and education programmes.

4.1 *Technical Advances*

The awareness of the importance of a well functioning traffic infrastructure in industrialised countries has resulted in the investment of new infrastructure. However, not only investment in the construction of new infrastructure is required.

During recent years the latter has been widely recognised in many countries and a large number of projects, sponsored by the European Union with the purpose of supporting the development of technical advances and approaches, has been initiated.

However, optimisation of O&M of structures and installations is not only a national and/or international obligation. The individual infrastructure owner shall initiate his own programmes of importance in his specific O&M optimisation process.

4.2 *Knowledge and Training*

The international knowledge within O&M is presently undertaking a very rapid development. It is the interest of all O&M infrastructure owners and operators to share this knowledge in order to protect their investments and carry out optimum O&M.

One of the most efficient ways is to work closely together with international contractors and consultants. This in combination with implementation of proper training and education programmes for his own employees allows an O&M optimisation to be carried out utilising the latest available knowledge at all times.

4.3 Manuals, procedures and instructions

The life of structures and some installations is several generations. The importance of preparing proper O&M manuals is particularly important in case of unique and major bridges and tunnels.

The latest developments in the information technology have made the argumentation for preparation of O&M manuals, procedures and instructions even stronger because of the development of intranet systems, which ensure very user-friendly systems, where the latest version is always available.

5. O&M ASSESSMENT AND REHABILITATION

5.1 Equipment and Systems

The utilisation of e.g. reliability centred maintenance techniques allows O&M assessment and rehabilitation to be performed as an overall optimisation process. This has resulted in significant saving for infrastructure owners and operators within the aeroplane industry, the offshore industry and the shipping industry.

The installation part of major/unique bridges and tunnels is significant irrespective of whether it is a road, a railway or a combined traffic corridor. The implementation of advanced assessment and rehabilitation methods in combination with a reliability centred O&M system has a significant optimisation potential.

5.2 Structures

Having survived the first and the second generation many of the larger structures, in addition to the first major rehabilitation, are facing a demand for a changed function because of a drastic increase in traffic volume, heavier and faster traffic, etc., which requires a significant reinvestment of the owner and/or operator and a potential major traffic disruption cost.

During the latest 5 years it has been documented that a complete verification of the actual loading on the structures, the actual material properties as constructed and the structural models, in combination with very efficient static and dynamic structural analysis edp-programmes, can be utilised to document higher allowable traffic loading than originally assumed.

The postponement of strengthening could e.g. in connection with major steel structures be based on the assumption that a special yearly inspection pro-

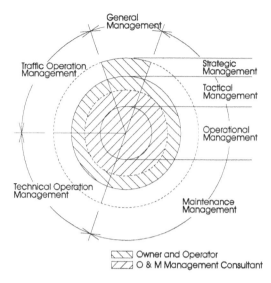

Figure 5. O&M management oganisation. Narrow combined owner and operator organisation.

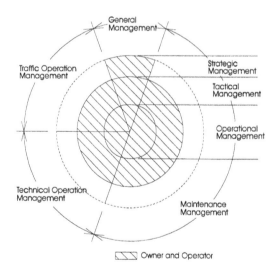

Figure 6. O&M management organisation. Large combined owner and operator organisation.

gramme is performed, in certain situations in combination with the implementation of a remotely controlled structural monitoring system.

6. O&M ORGANISATION

How to organise optimum operation and maintenance? There are different answers to this question in case of each unique infrastructure like The Great Belt Link, The Channel Tunnel, The Major State Owned Toll Bridges in California, etc. The answer i.a. depends on differences in ownership,

- Public
- Private
- Private companies backed with public guarantees,

differences in function,

- Road
- Rail
- Combined road and rail,

and differences in nationality,

- Danish project
- English - French project
- Californian project.

With reference to figure 4, figure 5 illustrates the O&M management organisation of a narrow combined infrastructure owner and operator organisation. The task of the owner and operator is limited to strategic management and tactical management in the form of contracting. The remaining management is outsourced to an O&M Management Consultant. The operational activities can be performed by the same Consultant in case he also has a proper technical background. Otherwise this work has to be contracted to Consulting Engineers thereby resulting in an increased management task. All work to be carried out by contractors is on a day to day basis managed by the O&M Management Consultant.

Figure 6 illustrates the similar O&M management organisation of a large combined infrastructure owner and operator organisation. In this case all strategic, tactical and operational management work is performed by the owner and operator. All operational work is assumed carried out by Consulting Engineers.

7. CONCLUSION

Optimisation of O&M of structures and installations in traffic corridors is a complex multidisciplinary effort, which at strategic level requires a management that not only focuses on the financial results of the owner/operator.

In all industrialised countries an increasing awareness of the relationship between financial results, environment, customer satisfaction and image in the society requires the attention of the O&M management in order to ensure a total optimisation.

It is not possible to outline a single parameter of the highest importance in the optimisation process. However, the utilisation of a well structured O&M management model with a clear definition of the various management tasks at strategic, tactical and operational levels is a first major step in an optimisation process.

If the latter is ensured in combination with the use of:

- efficient management methods and tools
- continuous O&M development of knowledge and competence
- the latest O&M assessment and rehabilitation approaches
- the building up of O&M organisation based on a well structured management model
- utilisation of the international capabilities of O&M Management Consultants, Consulting Engineers and Contractors as well as local skills,

optimisation of the O&M of structures and installations can be achieved.

Operation and Maintenance of Large Infrastructure Projects, Vincentsen & Jensen (eds)
© 1998 Taylor & Francis, ISBN 90 5410 963 7

Technical advances within maintenance and repair of structures

H.R.Sasse & P.Schiessl
University of Technology Aachen, Buildings Materials & Institute of Building Materials Research, Germany

ABSTRACT: In the past the chemical industry developed and offered products and techniques based on the material properties and provided the market in this way with recipies for maintenance and repair of concrete structures. The very complex deterioration mechanisms of concrete and steel in concrete have not sufficiently been taken into account. This situation in the past could be called the trial and error period. The main technical advance in the field in the past 10 years is the fact, that the civil engineering society learned and accepted that maintenance and repair is a real engineering task needing a profound knowledge based design and execution. This starts with the clear understanding of the mechanisms leading to deterioration, followed by strategies, concepts and procedures for maintenance and repair based upon the mechanisms and ends with assessment and monitoring procedures and techniques again based upon the mechanisms. That means that we nowadays solve the problem on a sound technical basis, design the maintenance and repair work and on that basis define the needed material properties for every specific maintenance or repair case. In this way the best technical and economical solution can be found. This recent change of the situation is documented in the RILEM Recommendation 124 SRC "Strategies for Repair of Concrete Structures Damaged by Steel Corrosion" and in the series of drafts of CEN-standards on protection and repair of concrete structures.

1 INTRODUCTION

Considerable technical advances have been gained in the area of maintenance and repair of structures within the last two decades. A short contribution like this paper can only highlight some aspects of the recent developments.

Maintenance and repair aspects within this paper are restricted to effects of environmental actions and ageing in service phenomena. Wear and fatigue effects will not be considered. Furthermore, main emphasis is laid on concrete structures, as maintenance and repair related to environmental actions is of higher actuality compared to steel structures - corrosion protection and maintenance of the protective systems have a long tradition and a high efficiency for steel structures.

2 CONCRETE STRUCTURES

2.1 *Identification of technical advances*

In the past the chemical industry developed and offered products and techniques based on the material properties and provided the market in this way with

recipies for maintenance and repair of concrete structures. The very complex deterioration mechanisms of concrete and steel in concrete have not sufficiently been taken into account. This situation in the past could be called the trial and error period.

The main technical advance in this field in the past 10 years is the fact, that the civil engineering society learned and accepted that maintenance and repair is a real engineering task needing a profound knowledge based design and execution. This starts with the clear understanding of the mechanisms leading to deterioration, followed by strategies, concepts and procedures for maintenance and repair based upon the mechanisms and ends with assessment and monitoring procedures and techniques again based upon the mechanisms. That means that we nowadays solve the problem on a sound technical basis, design the maintenance and repair work and on that basis define the needed material properties for every specific maintenance or repair case. In this way the best technical and economical solution can be found.

This recent change of the situation is documented in the RILEM Recommendation 124-SRC "Strategies for Repair of Concrete Structures Damaged by Steel Corrosion" (RILEM Technical Recommendation 124-SRC 1994) and in the series of drafts of CEN-standards on protection and repair of concrete structures.

2.2 Relevant degradation processes - Example corrosion of steel in concrete

Most of the degradation processes leading to concrete degradation or steel corrosion are fairly well understood nowadays. A presentation of the most important mechanisms on an engineering level as a basis for design for durability has been published by CEB Task Group on Durability (CEB Durability of Concrete Structures 1982, CEB Design Guide for Durable Concrete Structures 1992).

The new approach as indicated in chapter 2.1 will be demonstrated for one of the most important deterioration processes - corrosion of steel in concrete, being the deterioration mechanism causing by far the greatest problems during the service life of concrete structures - at least with respect to the overall costs to repair damages caused by environmental actions.

It is common knowledge that steel in well composed, placed, compacted and cured concrete is excellently and durably protected against corrosion by the alkalinity of the porewater, causing passivation of the steel surface. It is furthermore well known that basically two processes may lead to depassivation of steel in concrete

- carbonation of concrete
- chlorides above critical values.

As long as the "depassivation front" has not reached the reinforcement, corrosion protection is ensured. Only after penetration through the concrete cover carbonation or chlorides may initiate corrosion. This situation is normally sketched in a general deterioration scheme first published by Tuutti in 1982 (Tuutti 1982), see Figure 1.

Steel corrosion visible with the naked eye is the outward manifestation of the effects of a number of small corrosion cells on the steel surface.

In simplified terms a corrosion cell is essentially a short-circuited battery consisting of a metallically and electrolytically connected anode and cathode. Unlike the process in a rechargeable battery, the corrosion process is not, however, reversible.

The corrosion process can be subdivided into two sub-processes, as outlined in Figure 2:

- At the anode, in the presence of water iron ions pass into solution, separating from the electrons. They are converted into rust products in further reactions.
- At the cathode, electrons, water and oxygen are converted into hydroxyl ions. The cathodic process doesn't cause any deterioration of the steel.

These hydroxyl ions transport the negative charge in the electrolytic fluid in the concrete pore system through the electrical field created between the anode and the cathode, towards the direction of the anode. Near the anode, they react with the steel ions in solution. Depending on moisture and aeration conditions, this intermediate product may continue to react, producing the final corrosion products.

Individual processes are in fact much more complicated. The RILEM Report (RILEM Report of Technical Committee 60-CSC 1988) indicates the state of the knowledge in this field.

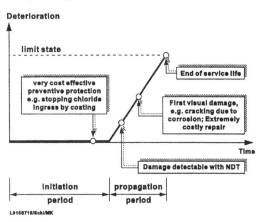

Deterioration

L9158715/Schi/MK

Figure 1. General deterioration scheme.

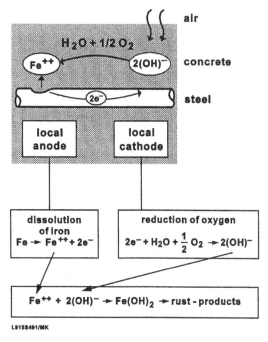

L9153491/MK

Figure 2. Corrosion of steel in concrete - scheme.

122

In order for the corrosion process to take place, a number of preconditions for the anodic and cathodic process and for the electrolytic process must be satisfied simultaneously:

- As already noted, there must be differences in potential. The preconditions for sufficiently large differences in potential are, however, virtually always met, and in the case of chloride-induced corrosion these may be, after local depassivation, several 100 mV.
- Anodic and cathodic surface zones of the steel must be connected electrically and electrolytically, i.e. a flow of electrons and ions between them must be possible.
- The metallic connection necessary for an electron flow from the anode to the cathode is provided by the reinforcement system in the reinforced concrete. The electrolytic connection is represented by the concrete. This must, however, be sufficiently moist, since otherwise there can be virtually no migration of ions. In dry interior situations, for example, the electrolytic conductivity of the concrete is too low to permit corrosion of the reinforcement, even if the carbonation front reaches the reinforcement, leading to loss of alkalinic protection.
- Anodic solution of iron must be possible due to depassivation of the steel surface. The cathodic process can, however, take place even in zones with a passive steel surface.
- Sufficient oxygen must be available at the cathode. There must be continuous diffusion of oxygen from the surface of the concrete to the steel surface acting as the cathode. There is therefore practically no risk of corrosion to reinforced steel components which are permanently immersed in deep water.

If all conditions for corrosion are fulfilled simultaneously, the reinforcement will corrode. If only one of the conditions can be eliminated, corrosion can be prevented or brought to a stop.

Starting from this knowledge maintenance and monitoring strategies and techniques (example see chapter 2.3) as well as principles and systems for repair have been developed (see chapter 2.4).

2.3 Consequences for maintenance and monitoring - Example corrosion of steel in concrete

As a consequence of the identified degradation process maintenance needs to be oriented towards extending the initiation period (Fig. 1) to the intended service life. This should be done by regular measurements and observation of the ingress of the depassivation front (chlorides or carbonation) to enable the provision of protective measures before deterioration starts. Protective measures may be

suitable surface protection systems or the installation of cathodic prevention systems. Examples are given in chapter 2.4.

Performance testing and checking the real quality of the finished structure as an important part of the quality assurance and quality control system is the starting point of any monitoring and maintenance procedure.

Jensen & Andersen (Jensen & Andersen 1994) have defined three levels of monitoring of structural performance.

Low   Periodic visual general inspections as integrated parts of maintenance management systems

Medium Periodic visual general inspections in combination with aperiodic and periodic special inspections using non-destructive and destructive testing techniques all as integrated parts of maintenance management systems

High   Systems for monitoring of structural performance implemented already during design and construction, periodic and aperiodic special inspections using non-destructive and destructive testing techniques and visual periodic inspections all as integrated parts of maintenance management systems.

2.3.1 Periodic visual general inspections

This type of monitoring is widely used nowadays. However, it is more trouble-shooting than real monitoring. Most of the environment related deterioration mechanisms show an initiation period before deterioration starts (Fig. 1).

For the deterioration process "corrosion of reinforcement in concrete" - causing by far most of the damages - the deterioration can be detected by visual inspection only at a rather late stage when cracking and spalling occurs. The cost for repair at this stage of damage is extremely high compared to preventive protection stopping the initiation process (carbonation or chloride ingress in case of reinforcement corrosion) before deterioration starts.

2.3.2 Special inspections using non-destructive and destructive techniques

Non-destructive testing techniques usually operate from the concrete surface and can detect ongoing deterioration before visual damage occurs (e.g. potential mapping). Therefore repair measures can be taken before the structural integrity has been weakened by e.g. cracking and spalling. Additionally, destructive testing can be performed (e.g. chloride profiles, moisture content of concrete) to get a better picture of the overall situation. However, all NDT and DT leaves a lot of uncertainty in diagnosis because of uncertainties in the techniques themselves combined with a lack of knowledge of structural performance in the past and a very limited possibility of predicting the future deterioration rates with different levels of maintenance work.

2.3.3 Continuous monitoring of structural performance with implemented sensors

The highest level of monitoring is the implementation of sensors during the construction process to continuously monitor the performance of the structures. Successfully used monitoring systems of this sort are

- vibrating wire sensors to monitor the time dependent development of stresses
- optical fibre sensors to monitor deformations and displacements (e.g. Miesseler & Lessing 1989)
- corrosion cells to monitor the progress of the depassivation front (see below)
- moisture sensors to monitor the conductivity of concrete and/or the efficiency of protective measures (Schiessl & Breit 1995).

The evident advantage of these monitoring techniques is that the development of the structural performance with time is continuously monitored and preventive protection or repair measures can be taken at comparatively low costs before real damage occurs and the structural integrity will be affected (see Fig. 1). Furthermore a prediction of future performance of the structure is possible at any time and

eases the decision on the adequate time of interventions.

2.3.4 Example: Macrocell corrosion sensors

The operation of a macrocell consisting of a piece of black steel (anode) and a noble metal (cathode) is shown in Figure 3. In chloride free and non-carbonated concrete, both electrodes are protected against corrosion due to the alkalinity of the pore solution of the concrete (passive state). The electrical current between both electrodes is negligibly low under such conditions. If, however, a critical chloride content is reached, or if the pH-value of the concrete decreases due to carbonation, the steel surface of the anode is no longer protected against corrosion, while the cathode remains passive. The local separation of anodically and cathodically acting areas leads to an electron flow between the black steel and the passive cathode if an electrical connection is provided. The flow can easily be measured at the external cable connection using a low resistance amperemeter.

Figure 3 shows the result of electrical current measurements (demonstration test in the laboratory) between a black steel anode and a stainless steel cathode in two concretes with different water-cement ratios. The macrocells were embedded with a concrete cover of only 5 mm (0.20 in.) to initiate corrosion by applying a chloride solution on the concrete surface within a short period of time. The results of the macrocell current measurements showed that the critical chloride content reached a depth of 5 mm (0.20 in.) at the specimen with w/c = 0.7 about 80 days after concrete placement. This caused a significant increase of the macrocell current while the specimen with a lower w/c and a higer resistance against chloride diffusion remained passive.

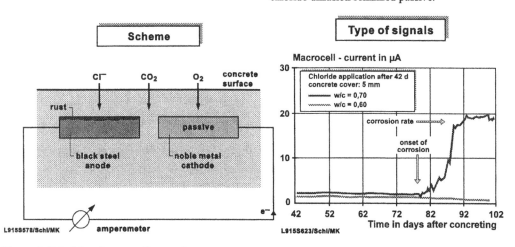

Figure 3. Principle of macrocell corrosion sensors.

Until now about 500 macrocell corrosion sensors, measuring the time dependent ingress of the depassivation front towards the reinforcement have been installed into various structures.

The system of the sensors used is sketched in Figure 4. Within one sensor element normally 6 single anodes are positioned between the concrete surface and the reinforcement e.g. at distances of 5 mm from each other. As soon as the depassivation front (e.g. critical chloride content for the given concrete in the given environment or carbonation) has reached an anode a corrosion current can be measured when the anode is coupled to an embedded cathode consisting of a noble metal. With time anodes at increasing depths will be activated thus giving information on the ingress of the depassivation front (see Fig. 4).

The layout of the sensor systems allows, besides the current readings other measurements improving the information on the overall risk within the monitored structure:

- The measurement of the potential between the anodes or the reinforcement and the noble cathode gives a further information on the onset of corrosion.
- The simultaneous measurement of the temperature by means of an incorporated temperature sensor allows a more detailed interpretation of the current readings.
- The anodes can be used as measuring electrodes for AC resistance measurements at different distances from the concrete cover. This type of readings for example can be used to monitor the efficiency of coatings, preventing water ingress into the concrete.

The advantage compared to e.g. taking chloride profiles is that the real corrosion initiating chloride levels are detected directly. By continuously measuring the progress of the depassivation front protective measures can be taken before it reaches the reinforcement. A detailed layout of the sensor system (anode part) is shown in Figure 5.

2.3.5 Installation of macrocell corrosion sensors into the structures of the Great Belt Link

All the structures of the Great Belt Link-Tunnel, Western Bridges, Eastern Bridge - have been instrumented with corrosion sensors. Of course not all single elements need to be instrumented, a representative number of specific elements (tunnel segments, pier shafts, bridge girders etc.) is sufficient. The basic strategy of high level monitoring (see above) is to put results of the continuous monitoring with sensors together with other representative results of NDT- and DT-results at regular inspection intervals and to draw conclusions.

- Tunnel -
Altogether 204 sets of sensors have been installed into the 7.9 km long twin tunnels between Zealand and Sprogø at the eastern end of the Great Belt. Water depths above the tunnel increases to a maximum of about 55 m.

84 of the prefabricated segments within 14 tunnel rings have been equiped with sensors. To monitor the corrosion risk for the outer layer (chloride ingress) and the inner layer of reinforcement (chloride accumulation due to evaporation and carbonation) anode ladders (Fig. 5) have been installed on both sides of the segments. All the cathodes have been positioned on the inner side of the tunnel, as insufficient oxygen supply for the cathodic reaction must be expected at the outer side.

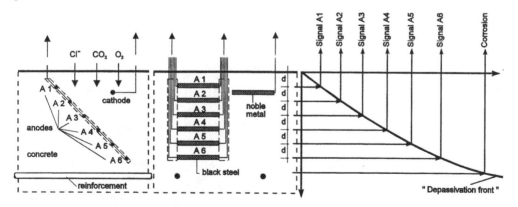

L915S576/Schl/MK

Figure 4. Scheme of a macrocell corrosion sensor monitoring the ingress of the depassivation front.

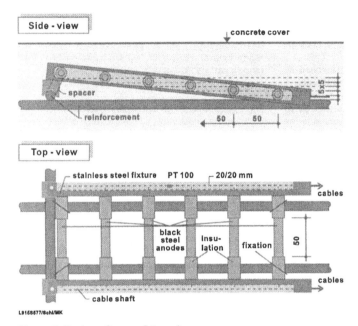

Figure 5. Design of a set of 6 anodes.

- Western Bridges -
6 pier shafts and girders out of 64 have been provided with 30 sets of sensors each (in total 180 sets of sensors). The sensors have been positioned in the most critical areas of the structure:

- construction joints between caissons and pier shafts
  (in-situ concrete connection under water)
- tidal and splash zones of the pier shafts (see Fig. 6)
- construction joints between casting sections of the precast girders
- salt spray zones of the girders
- parapets of the girders (deicing salt attack)

- Eastern Bridges -
Altogether 42 anode-ladders have been installed into selected positions of one pylon, one anchor block and one pier shaft. For the determination of the most critical areas the main wind- and water-flow-directions have been taken into account. 1/3 of the sensors have been installed into the first casting joints above the mean water level.

2.4 *The engineering approach to repair projects*

2.4.1 *General*

The phases of repair projects follow more or less a logical sequence, which is dominated by engineering aspects, but also includes commercial arguments and personal preferences. Figure 7 gives a general

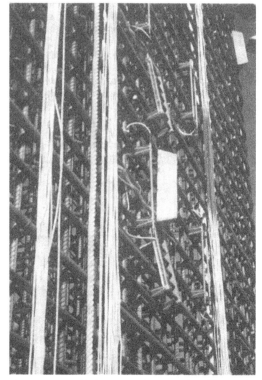

Figure 6. Installed sensors (6 sets of anodes and 1 cathode - Pier shaft of the Western Bridge, Great Belt Link).

126

MANAGE-MENT OF THE STRUCTURE	PROCESS OF ASSESSMENT	GENERAL PLANNING	DESIGN OF REPAIR WORK	REPAIR WORK	ACCEP-TANCE OF REPAIR WORK	MAINTE-NANCE
• Conditions and history of structure • Documentation of maintenance results in a management system	Assessment of defects, their classification and causes	Options Principles Methods	Definitions of the intended use of products Requirements - substrate - products - work - specifications - drawings	Choice and use of products and equipment. Tests for quality control. Health and safety	Acceptance testing Remedial works Documentation	Systematic inspections, evaluation of results, maintenance works

Figure 7. The phases of repair projects.

scheme. The following chapter is related to the new European Standard ENV 1504, Part 9: "Protection and Repair of Concrete Structures: General Principles for the Use of Products and Systems" (ENV 1504-9 1996) and gives background information for the normative rules.

2.4.2 *Process of assessment*

Assessment may be taken in several stages. The purpose of a preliminary stage is to advise on the immediate safety of the concrete structure, to give an informed opinion on the urgency of commissioning further surveys, protection or repairs, and to outline what testing should be done to establish the causes and likely extent of the defects.

The purpose of the main assessment is:

- to identify the cause or causes of defects
- to establish the extent of defects
- to establish whether defects can be expected to spread to parts of the structure that are at present unaffected
- to assess the effect of defects on structural safety
- to identify all locations where protection or repair may be needed.

Assessment should include testing or other investigation with the aim of revealing hidden defects and causes of potential defects. Due to the time-dependent deterioration processes previous assessments need to be updated.

The process of assessment generally includes:

- valuation of the original design approach
- conditions during construction (including climatic conditions
- the history of the concrete structure
- the conditions of use (e.g. loading)
- the environment, including exposure to contamination
- the present condition of the existing concrete structure, including non-visible and potential defects
- requirements for the future use of the concrete structure.

The nature and the causes of defects result from a great variety of parameters, as mentioned in Figure 8. The causes include:

- defects due to inadequate construction or materials (such as inadequate structural design, inadequate mix design, insufficient compaction, insufficient mixing, insufficient cover, insufficient or defective waterproofing, contamination, poor or reactive aggregates, and inadequate curing)
- defects revealed during service (such as foundation movement, impacted movement joints, overloading, impact damage, and expansion forces from fires)
- external environment and agents (such as severe climate, atmospheric pollution, chloride, carbon dioxide, aggressive chemicals, erosion, aggressive groundwater, seismic action, and electric stray currents.

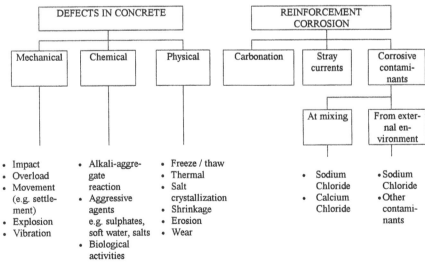

Figure 8. Common causes of defects.

As a general rule the initial causes of the defects need to be identified and corrected before a successful repair can be carried out. If correction is not possible, e.g. the use of de-icing salts, the repair must be designed to resist the cause as far as possible.

2.4.3 *Repair strategy*

The selection of a strategy for repair can never be taken on technical grounds alone. In addition to considering the technical possibilities, it is necessary to consider the economic circumstances which affect the decision. As much information as possible has to be provided to enable an informed judgement to be reached on the relative costs and benefits of the possible technical options for repair.

The design life of the repaired concrete structure is a key consideration in the choice of the repair method. Options range from those which can restore the design life of the concrete structure in a comprehensive single operation, to simpler options which may require repeated maintenance activities. Options with a long maintenance-free life are not necessarily preferable to those with a shorter life and the appropriate choice will depend on the circumstances of the individual case. There is normally a wide range of options, such as

- do nothing for a certain time
- re-analysis of structural capacity, possibly leading to downgrading of the function of the concrete structure
- prevention or reduction of further deterioration, without improvement of the concrete structure
- improvement, strengthening or refurbishment of all or part of the concrete structure
- reconstruction of part or all of the concrete structure
- demolition of all or part of the concrete structure.

Consideration of the options and their consequences will generally include examination of different aspects, for example levels of initial cost, maintenance costs and the possible need to introduce restrictions in the use of the structure. A different level of risk of future deterioration is likely to be associated with each option. The choice of appropriate action needs also an objective analysis and evaluation of the principles and methods which are described in the following.

2.4.4 *Principles of repair and protection*

The principles in Tables 1-2 are based on the chemical or physical laws which allow prevention or stabilisation of the chemical and physical deterioration

processes in the concrete or the electrochemical corrosion processes on the steel surface (ENV 1504-9 1996).

Methods in Tables 1-2 which make use of the principles are given as examples: other methods may be used if it can be shown that they comply with the principle.

Several methods may be used in combination, but of course the possibility of undesirable as well as desirable interactions between them have to be considered. The influence which repair of the concrete may have on the corrosion of reinforcement, for example by entrapping moisture or by raising the ambient temperature, is also part of the consideration to be made.

Examples for undesirable consequences connected with particular methods may be:

- reducing the moisture content, which may increase the rate of carbonation
- surface coating, which may entrap moisture and may destroy adhesion or reduce frost resistance
- electrochemical methods, which may cause embrittlement of susceptible prestressing steel, alkali aggregate reaction with susceptible aggregates, a decrease in frost resistance due to increased moisture contents, or, if under water, corrosion in adjacent structures or vessels
- limiting oxygen content by surface coating or saturation, which can cause increased corrosion if reinforcement in the protected zone is electrically connected to reinforcement in any unprotected zone.

It is self-evident that products and systems should be compatible with each other and with the original concrete structure.

Where there is a history or risk of reinforcement corrosion, it is necessary to consider principles 7 to 11 in addition to principles 1 to 6.

2.4.5 *Selection of appropriate products*

Only after assessment of the status of the structure, after elaborating the repair strategy and after defining the principles and methods of repair the basis is given to define characteristics and requirements for materials to be used.

After these definitions it is up to the design engineer to select commercial products which meet theses requirements.

In future, products and systems which have appropriate properties will be described in the EN 1504 series or other relevant EN or European Technical Approvals.

Some parts of the repaired concrete structure may have an expected service life which is short compared with that of the rest of the concrete structure. Familiar examples are surface coatings, sealant, and weather proofing materials. If the integrity of the protection or repair depends on such parts, it is essential that they be regularly inspected, tested and renewed if necessary. In these cases maintenance should include:

- a precise time schedule for inspections
- the system of inspection which is to be used, how results are to be recorded
- a specification for continuous treatment (if any is required), for example cathodic protection or monitoring equipment
- a statement of precautions to be taken or prohibitions to be enforced, for example maintenance of surface water drainage, maximum pressure for washing or prohibition of the use of de-icing salt.

A maintenance management system should be im-

Table 1. Principles and methods related to defects in concrete.

Principle No	Principle and its definition	Methods based on the principle
Principle 1 [PI]	Protection against ingress Reducing or preventing the ingress of adverse agents, e.g. water, other liquids, vapour, gas, chemicals and biological agents.	1.1 Impregnation Applying liquid products which penetrate the concrete and block the pore system. 1.2 Surface coating with and without crack bridging ability 1.3 Locally bandaged cracks 1.4 Filling Cracks 1.5 Transferring cracks into joints 1.6 Erecting external panels 1.7 Applying membranes
Principle 2 [MC]	Moisture Control Adjusting and maintaining the moisture content in the concrete within a specified range of values.	2.1 Hydrophobic impregnation 2.2 Surface coating 2.3 Sheltering or overcladding 2.4 Electrochemical treatment Applying a potential difference across parts of the concrete to assist or resist the passage of water through the concrete. (Not for reinforced concrete without assessment of the risk of inducing corrosion).
Principle 3 [CR]	Concrete Restoration Restoring the original concrete of an element of the structure to the originally specified shape and function. Restoring the concrete structure by replacing part of it.	3.1 Applying mortar by hand 3.2 Recasting with concrete 3.3 Spraying concrete or mortar 3.4 Replacing elements
Principle 4 [SS]	Structural Strengthening Increasing or restoring the structural load bearing capacity of an element of the concrete structure.	4.1 Adding or replacing embedded or external reinforcing steel bars 4.2 Installing bonded rebars in preformed or drilled holes in the concrete 4.3 Plate bonding 4.4 Adding mortar or concrete 4.5 Injecting cracks, voids or interstices 4.6 Filling cracks, voids or interstices 4.7 Prestressing - (post tensioning)
Principle 5 [PR]	Physical resistance Increasing resistance to physical or mechanical attack.	5.1 Overlays or coatings 5.2 Impregnation
Principle 6 [RC]	Resistance to chemicals Increasing resistance of the concrete surface to deteriorations by chemical attack	6.1 Overlays or coatings 6.2 Impregnations

Table 2. Principles and methods related to reinforcement corrosion.

Principle No	Principle and its definition	Methods based on the principle
Principle 7 [RP]	Preserving or Restoring Passivity Creating chemical conditions in which the surface of the reinforcement is maintained in or is returned to a passive condition.	7.1 Increasing cover to reinforcement with additional cementitious mortar or concrete 7.2 Replacing contaminated or carbonated concrete 7.3 Electrochemical realkalisation of carbonated concrete 7.4 Realkalisation of carbonated concrete by diffusion 7.5 Electrochemical chloride extraction
Principle 8 [IR]	Increasing Resistivity Increasing the electrical resistivity of the concrete.	8.1 Limiting moisture content by surface treatments, coatings or sheltering
Principle 9 [CC]	Cathodic Control Creating conditions in which potentially cathodic areas of reinforcement are unable to drive an anodic reaction.	9.1 Limiting oxygen content (at the cathode) by saturation or surface coating
Principle 10 [CP]	Cathodic Protection	10.1 Applying electrical potential
Principle 11 [CA]	Control of Anodic areas Creating conditions in which potentially anodic areas of reinforcement are unable to take part in the corrosion reaction.	11.1 Painting reinforcement with coatings containing active pigments 11.2 Painting reinforcement with barrier coatings 11.3 Applying inhibitors to the concrete

plemented to ensure that the required maintenance is carried out.

3 STEEL STRUCTURES

Whereas the need of maintenance and eventually special protection of concrete structures has been only realised and accepted during the last 25 years, the protection of steel structures against environmental actions, i.e. mainly corrosion, was an accepted part of design and maintenance of steel structures since ever. Therefore maintenance and renewal of corrosion protection systems are a high technical level since long time ago.

Technical advances have been gained in the last years in another area of steel structures: design and verification methods for the residual safety, remaining service life and rehabilitation of old steel bridges.

A considerable part of existing road and railway steel bridges have been built more than 6 to 8 decades ago. Many of these old bridges have undergone various phases of repair or strengthening after damages in the world war or due to changes of the service requirements. For these bridges the question of the given safety for the actual traffic loads and the remaining service life need to be answered.

Similar to the reported situation for concrete structures, increased knowledge about material degradation has enabled new approaches to the above mentioned areas. Sedlacek and Hensen (Sedlacek & Hensen 1992) have presented new methods for the identification and thoughness related material checks of old steel bridges that allow to get a complete view on the residual safety and service life of such bridges and also permit to determine measures for strength and toughness related strengthening.

The fracture mechanic based verification procedure have been simplified and presented in such a form, that the assessements can be carried out as easily as the conventional strength verifications.

Similar approaches, using the path independent J-integral and its value at incipient crack extension have been presented by K. Eriksson (Eriksson 1994) to assess old steel bridges in Sweden. Brühwiler and others (Brühwiler et al. 1989) have shown how calculation models based on fracture mechanisms can be applied to assess critical crack lengths in structural elements.

4 CONCLUSIONS

Recent developments in maintenance and repair of both concrete and steel structures have clearly demonstrated that a profound knowledge of degradation mechanism on a material science level and its transfer to practical application on an engineering level are important driving forces for technical advances in design, execution, maintenance and repair of structures.

5 REFERENCES

Brühwiler, E. & Hirt, M.A. & Morf, U. & Huwiler, R. 1989. Bewertung der Spontanbruchgefahr angerissener Brückenbauteile aus Schweißeisen. Wilhelm Ernst & Sohn Verlag für Architektur und technische Wissenschaften, Berlin, Germany. *Stahlbau 58*, No. 1, pp. 9-16

CEB / Comite Euro-International du Beton 1982. Durability of Concrete Structures : State of the Art Report. *Bulletin d'Information*, No. 148

CEB / Comite Euro-International du Beton 1992. Durable Concrete Structures : Design Guide. *Bulletin d'Information*, No. 183

ENV 1504-9 1996. Products and Systems for the Protection and Repair of Concrete Structures - Definitions, Requirements, Quality Control and Evaluation - Part 9: General Principles for the Use of Products and Systems

Eriksson, K. 1994. A Fracture Mechanics Assessment Method for Older Structural Steels. *Proceedings of the 10th Biennial European Conference on Fracture "Structural Integrity: Experiments, Models, Applications* (ECF 10), Berlin, Germany. Volume I, pp. 561-570

Jensen, J.S. & Andersen, E.Y. 1994. Monitoring of Structural Performance - A Must for the Future. *Proceedings of the International Conference "Concrete Across Borders"*, Odense, Denmark, pp. 585-594

Miesseler, H.-J. & Lessing, R. 1989. Monitoring of Load Bearing Structures With Optical Fiber Sensors. *IABSE*, Lisbon

RILEM (edited by P. Schiessl) 1988. Corrosion of Steel in Concrete. *Report of the RILEM Technical Committee 60-CSC*

Schiessl, P. & Breit, W. 1995. Monitoring of the Effectiveness of Surface Protection Systems After Repair Measures Using Multi-Ring-Electrodes. *International Symposium on Non-Destructive Testing in Civil Engineering* (NDT-CE), Berlin, Germany

Sedlacek, G. & Hensen, W. 1992. New Design Methods for the Rehabilitation of Old Steel Bridges. *Proceedings of the 3rd International Workshop on Bridge Rehabilitation*, Darmstadt, Germany, pp. 301-317

Tuutti, K. 1982. Corrosion of Steel in Concrete. *CBI Forskning Research*, No. 4

Operation and Maintenance of Large Infrastructure Projects, Vincentsen & Jensen (eds)
© 1998 Taylor & Francis, ISBN 90 5410 963 7

Bridge maintenance training: Experience and future trends

Anders Huvstig
Swedish National Road Administration (SNRA), Western Region, Göteborg, Sweden

ABSTRACT: This paper presents some ways of getting more experienced, understanding and creative people, who are working with bridge maintenance. This paper also describes examples of co-operations and systems, which are in use in order to collect and spread knowledge. Furthermore the paper will describe different ways of organising the work of a client, in order to exchange experiences and train the bridge maintenance people in Sweden. At last, this paper describes future trends, new ways of co-operation and the clients organisations in order to get more skilful and creative people in the future, who are going to work with bridge maintenance. What is described as a good way of working with maintenance of bridges does not necessary means that that the Swedish National Road Administration (SNRA) always works in this way.

1. INTRODUCTION

There are a great need for cost-effective strategies for maintenance and repair of bridges. One important part is to describe the actual condition, for example bearing capacity or degree of corrosion. Furthermore there is a need for understanding of the ageing mechanisms, which are active, and at which speed the process of ageing is going on in the actual environment. After this, you must connect these calculations and knowledge with a calculation of the whole life costing for reparation at different years. If you only have a limit amount of money, you also have to make a judgement of the lowest agreeable functional characteristics, which can be permitted, for the bridge. The mentioned processes can be a good ground for choice of strategy for maintenance and repair of bridges.

The most important part in these processes and systems are the people, who are working with maintenance and repair of bridges. These people must have very much of experience and knowledge from maintenance and repair of bridges. They also must have an understanding in the theories behind deterioration of different materials in a bridge, and creativity in order to make the right judgement in different situations.

2. MAINTENANCE

2.1 *Inspection*

Every bridge is normally inspected at certain intervals, normally 2-3 years in most countries.

The primary function of the inspection is to have a safe traffic on the bridge, and to get as low maintenance and society cost as possible.

Different bridges have a lot of different design, constructive parts and materials. There are also surroundings like the road, drainage and water etc., which can influence the bridge structure.

People who are responsible for the inspections must also be responsible for analysis of the bearing capacity of the bridges and give recommendations for necessary immediate repair.

Experienced bridge inspectors have a great knowledge in the history of a bridge and also in the development of deterioration. Therefore the bridge inspector ought to be involved in the work with constructive design for repair of bridges.

Depending on all these factors, a bridge inspector must have a large competence. Experience from bridge inspections, and a good education as a ground, is necessary. He or she must also have knowledge about all different materials and constructive parts of different bridge structures. Furthermore, an inspector must have so much knowledge in design technique and deterioration mechanisms that he can give recommendations, which are correct, and suggest creative solutions on difficult repair problems.

A bridge inspector also ought to have social competence, an ability to explain his ideas in such a way that other people understand them.

2.2 Continuos maintenance

Contractors in Sweden do all maintenance work on the road network today, in competition ("own production" in SNRA is here a contractor).

The contractor, who is responsible for an area of the road network, can do some work on bridges like weekly inspections in order to look for damages, if water is standing on the road surface etc. Twice a year he can also look on the function of the joints and bearings etc. Furthermore he can do the yearly cleaning every spring and repair work on pavement, railing and painting etc.

The people from the client are not directly responsible for the daily bridge maintenance, but if a contractor have to make a repair on the concrete or steel structures, he have to discuss this with the client's responsible bridge people.

People, who are responsible for the daily inspections and maintenance, must have a certain degree of knowledge in bridge technique.

People, who are doing repair work, ought to be certified for this kind of work, which means that they must have education in bridge technique and experience in the work with bridge repair.

2.3 Repair and strengthening

Before repair of concrete structures, there ought to do a thorough investigation in advance of things like carbonation of the concrete and chloride content in the concrete etc. Repair of structures with prestressed concrete is more complicated, and these structures need extra investigations. It is easier to investigate and repair the steel structures, but for an old bridge, the ability to weld the steel must be examined.

Strengthening of structures is first a question of whether or not strengthening is necessary, and if so, to choose the most suitable method for the strengthening. Already in this stage, you ought to engage an experienced design consultant.

For the work with design of strengthening or repair, you need an experienced design consultant. It is very important that the consultant is experienced and creative, because in strengthening and repair work there are a lot of possible design solutions. It is also easy to miss something in the design work, which will cost money during the construction phase, or give a bad quality of the work done. During this phase you also need experience from the construction phase.

When you are making repair work with old bridges on site, it often arises a lot of problems, and new weak points are detected. Therefore it is important that the people from the contractor, who is doing the work, have experience from this kind of work. Furthermore the client ought to have someone, who is available for the working site, and who has a large experience in this kind of work. It is important that the contractor and the client have a constructive and positive dialogue on the site in order to solve all technical problems.

2.4 Repair or renewal

In some occasions, the choice has to be made between repairing or strengthening a structure and a new bridge. This choice is dictated by numerous factors such as investment costs for the various alternatives, residual service life without action, interest rates, anticipated service life for a new bridge, financial prerequisites, etc.

For these decisions you need drafts of different design solutions in order to make calculations of the whole life costing for the alternatives.

2.5 Special bridges

Complicated bridges, like suspension bridges, cable staged bridges and movable bridges need a special attention. For inspection and repair you need to use certain specialists for steel, X-ray tests, cables, electricity, electronic equipment, machinery and the hydraulic system etc.

One possibility, which gives a continuity, is to engage these specialist for a long time as special persons at a consultant. If these specialists can have a continuous dialogue with the responsible people from the client, this will also improve the knowledge of the people from the client in order to be a better buyer.

2.6 *Maintenance management*

Records for bridges, like drawings, design calculations, inspection documents, repair work and strengthening have since a very long time been collected in archives.

Large owners of bridges have also had bridge maintenance departments in the organisation.

Today many countries are building bridge management systems, BMS, in order to store bridge data. These systems are also so called expert systems, with calculation modules, which are a help for calculating of the load bearing capacity, planning of maintenance and major actions, prices and in making result analysis etc.

When you use BMS in a country, it is necessary that the people, who are working with bridge maintenance, are accustomed to work with computer programs.

COMPETENCE

* KNOWLEDGE   * ABILITY TO SELL

* UNDERSTANDING   * CO-OPERATION

* CREATIVITY   * NO PRESTIGE

3. COMPETENCE

3.1 General

All the people, who are working with different kinds of bridge maintenance, must have competence and knowledge in many subjects.

First of all they must have a solid background in bridge technique. They ought to have worked with all phases in the life of a bridge, like planning, design, construction, maintenance, inspection, repair and strengthening etc. They also ought to have an education as at least engineer in civil engineering.

A good quality for an engineer is the ability to sell good ideas. It doesn't matter how well a proposed suggest is, if no one is using it.

Competence is also the willingness of doing a good work and accepting new ideas without prestige. This is not only a personal quality; it is also a question about the culture in the organisation.

Another important quality for a bridge engineer is their ability to co-operate, especially because the work with repair and strengthening is very much a teamwork.

A maintenance engineer of today, working for a client, must also be able to work with the computer in the daily work, and use all the present facilities in a BMS program.

Depending on what field of maintenance a bridge engineer is working with, he also needs special competence, se 3.3 to 3.6.

3.2 Creativity

One of the most important competence factors for people, who are working with bridge maintenance, is their creativity. A bridge maintenance engineer has to take a lot of decisions, where it is possible to choose many different solutions. If it shall be possible to choose the best solution, you must find all possible alternatives, and estimate the costs and quality (whole life costing).

It is possible to develop people's creativity. One very important part in the creative thinking is that you not only have knowledge about theories and deterioration mechanisms, you must also understand the background, why things happen, because the prerequisites are almost always different for different bridge repairs.

3.3 *Bridge inspector*

A bridge inspector needs knowledge about bridge design for the whole bridge structure and for the different parts of the bridge, material technique, the theories behind material deterioration and design reason for deterioration (cracks) etc. He also has to have knowledge in geotechnique, road technique and pavement technique etc.

A bridge inspector ought to be certified with special demands on education and experience from inspection, design and work on site with bridge repair (for example an inspector in USA need 10 years practice with inspections).

3.4 *Bridge maintenance engineer*

A bridge maintenance engineer needs a good knowledge in design and material technique, especially a good knowledge about the deterioration mechanisms.

It is of advantage if the work as a bridge maintenance engineer is combined with the work as a bridge inspector, because this gives them a broader experience. In any case, the bridge maintenance engineer ought to have worked with bridge inspections, and have at least the same competence as a bridge inspector.

3.5 *Bridge designer*

A bridge designer needs experience from the work with repair and strengthening of bridges. For this kind of design work it is especially important with experience and creativity, because you always has to create special design solutions.

A bridge designer must also have a lot of knowledge about material characteristics and deterioration mechanisms.

A bridge designer ought to be certified for bridge repair and strengthening, with special demands on education and experience (for example 10 years practice with design and an education in material technique and deterioration mechanisms etc.).

3.6 *Bridge site manager and labour*

A site manager for bridge repair and strengthening needs a broad experience from this kind of work. He also needs the same competence in concrete technology as a site manager for construction of new bridges.

Also the labour people ought to have experience from bridge repair.

4. THE WORK WITH KNOWLEDGE

4.1 *General*

Knowledge is one of the most important parts of the competence.

There is a lot of knowledge about bridges, in different organisations and also in different countries, knowledge about different problems with design and deterioration mechanisms for materials, which are used in bridges. If the people, who are working with bridges, use all this knowledge, it is possible to save a tremendous amount of money.

- In order to make it possible to work with the experiences from bridge construction and bridge maintenance, it is necessary to collect and store these data's in a systematic way. In a road administration a Bridge Management System is the new way of collecting and storing data.
- You must study and work up the stored data in order to investigate trends, and check if the theories, which you are using, are correct. This work ought to be done by experienced experts, sometimes in a co-operation with universities and research institutes.
- It is important to spread the collected knowledge inside an organisation in such a way that it is used in the daily work with repair and maintenance of bridges. It is also important to spread all this knowledge between clients, consultants, contractors and universities, and furthermore to spread knowledge between countries. This can be done with help of information, education, seminars, work shops, congresses and co-operation etc.
- New technique, and results from research and development, is also a kind of knowledge. It is important that the implementation of these results is done in a systematic way in order to get a good result. A good system for implementation can be grounded on a co-operation between one client and the university with help of consultants and contractors.

4.2 *Knowledge in an organisation*

It is important to collect experiences from the design, construction and maintenance of bridges. This collection work must be systematic and effective if it shall be valuable. It is also important to store the collected data in such a way that they are easy to find for everyone in the organisation, who needs the information.

Someone in the organisation ought to work with the collected data in a certain knowledge area, and be responsible for the reliability of the result from this work. He or she also ought to be responsible for spreading the knowledge in the organisation in the best way (written and/or oral information, seminars and workshops etc.).

If an organisation has many regional units, this organisation ought to have a network of people from the different regional units, for certain knowledge areas. This network can spread knowledge and discuss technical questions in order to give suggested recommendations. The leader of such a network ought to work on the headquarter, but there can be exceptions if some regional unit has a qualified expert.

4.3 *Knowledge exchange between organisations*

A lot of different organisations are working with bridge maintenance, and they all have different experiences, knowledge and competence. Different knowledge leads to misunderstandings and mistakes and a lack of knowledge is even worse. Mistakes, which are done depending on this fact, will cost much money for the society. The organisations, who are working with bridge maintenance, are clients, design consultants, contractors, technical universities and research institutes. All these actors ought to have a co-operation in some way in order to create a better result and making fewer mistakes.

Examples of possible ways of co-operation are given in chapter seven.

4.4 *Exchange of knowledge in the world*

There is a lot of knowledge in different countries in the world, but it is very difficult to spread this knowledge depending on different languages and rules, long distances and comparable few contacts.

Contacts can be established through organisations like PIARC, IABSE and NVF (the Nordic countries), but the contact persons are relative few. Another way of spreading knowledge is through international congresses. A third way of getting knowledge is through study visits and personal contacts.

One large problem is to find the knowledge in all the different countries. Another large problem is to spread the collected knowledge from the source to the people who shall use the new technique. The technique shall not only be spread, the users must

also accept for them new technique in a positive way. Most countries have no good solution on these problems today, but some suggestions are given in chapter seven.

4.5 *Knowledge from research and development*

There are a lot of new knowledge in the technical universities and research institutes. One problem is to implement this, often very theoretical, knowledge to the practical world of the bridge maintenance engineers. It is also difficult to implement new technique in the practical work. One reason for this can be that it is difficult for theoretical people and practical people to understand each other. Another problem is conservatism and lack of self-assurance, The best way of solving this problem is more co-operations between the clients bridge department and certain institutions on a technical university.

4.6 *Problems to overrun*

Some of the problems with collecting, storing and spreading experiences and knowledge are:

- There are many people who shall have the information, and just a few people who give it.
- There are very many sources for the knowledge. This makes the knowledge difficult to survey and sometimes are the results contradictionary.
- Many people do not accept new technique, depending on prestige, bad experiences and conservatism. Sometimes they are not able to understand the advantages with the new technique.

Ways of solving these problems are co-operation, education and a progressive culture in the organisation.

5. TRAINING

5.1 *General*

One important meaning with training is to give knowledge to the people, but there are also other important things, which can be trained, in order to improve the competence, like creativity, ability to sell good ideas, interest for new ideas without prestige and ability to co-operate.

This paper will mostly describe the training from a client point of view, but some parts of this training can also be used of consultants and contractors.

In the work with bridge repair and strengthening, it is important that contractors and consultants use some untrained people in order to give them experience in this kind of work.

One very good way of training is to work in co-operation with other actors on the market. Possibilities to start up a co-operation between client, contractors, consultants and universities of technology are described in chapter seven.

5.2 *Employing*

Some clients have maintenance production, bridge repair, bridge construction and design department in their own administration. These clients can employ young people with a new examination direct from the school, and give them the necessary experience inside their own organisation.

Today it is a trend for clients to have less own production. Instead they are buying planning and design from consultants, and construction, maintenance and repair from contractors. Therefore it is difficult, but not impossible, to give young engineers a necessary experience in the own organisation. In Sweden the SNRA has divided the administration into one part, which has the role of the client, and one part, which is responsible for all the own production.

It is important to have experienced people, who are working with maintenance. You can employ people, who have all the experience and competence that they need for the work. If so, they can start to work at once. However, it is difficult to find these people, especially with experience from bridge inspections. One good way is to employ people with experience from design work and bridge construction, and give them necessary experience from inspection and maintenance.

5.3 *Work rotation*

Work rotation is one of the most important parts of the education. Someone in the organisation ought to be responsible for that the work rotation is fulfilled in a good way.

In a work rotation the new employed must not work as an extra person beside the other people. He or she ought to have a job with own responsibility. In a too short time, as one or two month, it is difficult to have a job with responsibility. A suitable work is to follow a whole project, like the design of one bridge or bridge repair and also the construction of one bridge or one bridge repair.

One common problem is that a work rotation can be stopped too early, depending on that the person becomes "necessary to have" on one special department. One reason for this can be lack of people and another reason is that the new employed is very good in his work. Neither of these reasons is good for the whole organisation. For the future work, it is mostly better to fulfil the work rotation.

If the client have no own production, it is a problem to make a plan for a work rotation. For planning and design it is possible to solve this problem by lending people to a consultant, who is working with a project for the client. For construction work, the new employed can work with control on site for the client. It is difficult to be lent to a contractor depending on the economic discussions between client and contractor.

5.4 *Education in the Swedish National Road Administration (SNRA)*

The only formal demand on a bridge inspector in SNRA, is that the inspector shall have an about two week long course in bridge inspection.

The informal demands for a bridge inspector also ought to include:

- A two week long class I course in concrete technology.
- A one week long TR-steel course.
- A course in durability of concrete structures.
- Computer education.
- 2 years experience from bridge construction and bridge repair.
- Experience from bridge design.
- One or two years experience from bridge inspection (without having the responsibility).

Maintenance engineers need the same education and experience, but also a course in bridge repair.

Furthermore he or she need a qualified extra education in design, material technique and the theories behind material deterioration.

5.5 *Personal development in a regional road administration*

Besides the long courses, which are necessary to have, there is also a need for continuous education of all people, who are working with maintenance of bridges, in all the subjects mentioned in chapter two and three. Some examples from the SNRA, Western Region, are given here.

- A design course, with one of the most experienced designer in Gothenburg as a teacher, was given about ten times, one half day each third week. The meaning with this course was to give all people, who are working with bridges and geotechnique, knowledge about how different bridge types and parts of a bridge are working from the design point of view.
- A one day long course about communication, behaviour and relations between people has been arranged.
- The bridge maintenance people have done study visits, about once a year.
- Once a month, there has been a half day meeting between all people on the engineering department, who are working with bridges. On this meeting the most skilled people on certain subjects have informed the other about his speciality. Further there has been technical discussions about special actual technical themes, actual projects, training, employment, flexibility in the work, work rotation, how to support project leaders,

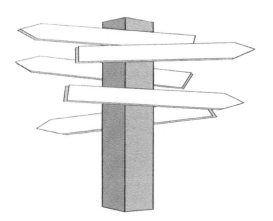

how to spread the knowledge and if there is a need of research and development in some area.

6. TRENDS IN THE WORLD

6.1 *General*

It is a trend to have more of the market economy also for a government, which means that it shall be a competition between own production (design, construction and maintenance), in an administration and the same production on the free market. In the end of this process, the own production is working as a "company", and the government can sell it out. One result of this is that an administration can concentrate on its role as a client.

It is also a trend to give the contractors more responsibility and let them do more work on the bridges. In order to manage this, clients and contractors are working with quality assurance systems and different kinds of contracts with functional demands. The consultants are also working with quality assurance systems in order to get better quality on the planning and design.

Another trend is that a road administration more and more is looking for the all the society costs during the whole lifetime of a project.

All actors on the market use more and more computer programs as help for the management. Almost every country has developed or is developing a bridge management system (BMS) today.

In the technical development, the fibre optic can perhaps change the inspection work. Fibre optic cables in a bridge structure can give a lot of valuable information about the condition of the bridge. Other technical developments, which can change the work with repair of bridges, are implementation of high performance concrete and concrete that doesn't need vibration.

6.2 *Functional demands*

Contracts with functional demands are more common today. Total contracts, with functional demands, where the contractor is responsible for the design and has a long guarantee time, is the best way of handling this.

One problem with very long guarantee times is that you don't know if the contractor will exist after so long time. It is also difficult to predict many functions in the future, but in a near future it is probably possible to solve many of these problems for concrete structures.

Functional demands gives the advantage for a client that he must know what function he is asking for, and not only ask for figures in a national code for bridges.

For the contractor this trend means that he has to improve his competence in design and durability questions.

6.3 *Quality assurance*

In many countries the contractors, and also the consultants, are using quality assurance programs in order to control their own work. If this shall give a reduction of the total cost, the client must reduce his own control and just do spot-checking.

The Swedish National Road Administration was probably the first road administration in the world, which put a demand on the contractors to have a Quality Assurance Program in 1988. The result from this work is up to now good. The contractors have been more aware of the quality of the product they deliver.

Today we are working with a system where the client not only are looking for the lowest price on a contract, but also the contractors ability and intention to work with quality, traffic safety and environment etc. This system, with so called soft factors for valuing the contractors, must complete the quality assurance system, if the contractors shall have an economical reason to work with quality assurance in a serious way.

For the contractors and consultants, who are working with quality assurance systems, this means that they are taking a greater responsibility and risks, but it also gives possibilities to create better technical solutions, and earn money on creativity and competence.

6.4 *Whole Life Costing (WLC)*

When you are buying a car or a house privately, you are not buying the cheapest product. Instead you try to calculate all your costs during the whole lifetime, in order to choose the alternative, which gives you the lowest cost.

Governments in different countries are trying to do the same. Therefore they are developing calculation models in order to calculate the whole life costing (WLC) and also the environmental impact (Life Cycle Analysis, LCA), so it is possible to compare different constructive solutions.

6.5 *BMS*

Many countries are working with bridge management systems in order to do the right maintenance, repair and rehabilitation work etc. on the bridges. In the future, many of the computer systems, which are used by consultants and contractors have to give information to the BMS. Perhaps, it is also possible for them to get information and knowledge back from BMS.

7. CO-OPERATION

7.1 *General*

With own departments for design, construction, repair and maintenance, a normal road administration of yesterday had enough with competence and experience about bridges. In this situation there was not so much need for co-operation. Most of the knowledge where spread from the head quarter with help of the bridge code and other instructions or from the experts.

Tomorrow there will be many actors, who are working with, and have competence in, bridge maintenance. There are a lot of consultants for the design and many contractors for repair and construction. This makes it more difficult to spread the knowledge.

Co-operation is a very good competence training for individual people, and co-operation is also a good way of spreading knowledge.

7.2 *Future development*

With all the actors on the market, who are working with bridges, there is a strong need of co-operation in order to collect experiences, spread the knowledge, discuss technical problems, give recommendations, suggest common rules, implement new technique and give ideas for research and development.

In order to initiate research and development and implement the results in form of new technique, there ought to be a close co-operation between clients and the technical universities, especially on a regional level. An even better way of co-operation is if consultants and contractors take part in this.

Depending on the quality assurance, there is a greater need of co-operation between client and contractor on a working site. A good example on this is when the representatives for the client and the contractor have a fruitful dialogue about the technical questions, and together try to foresee and avoid deviations and defects. There can also be a co-operation between client and contractor about other subjects on a bridge construction site, like technical development.

Experiences in the past were only collected from some of the projects. BMS gives a new possibility to collect experiences from the different actors. BMS has also a possibility to be a central storage for knowledge about bridges, where also consultants and contractors can get actual information.

7.3 *Examples from the work in the Western Region of SNRA*

An example from the Western Region of SNRA is a broad co-operation between the road and railway administrations, Gothenburg town, all contractors, all consultants and Chalmers University of Technology. This is a loose co-operation, who arranges education for bridge labours and young designers, discuss questions about bridge technique, initiates research and development, arranges a yearly bridge builders day with about 400 participants from the Scandinavian countries every year and other things of interest in connection with bridges.

Another example is that the Region has signed an agreement with Chalmers University of Technology in order to get a close co-operation between client and university on a regional level. One of the main reasons for this is to get a dialogue between people from the university and people from the Western Region. Another reason is to implement new technique in a scientific way. Today we are working together in several projects, and one of these is implementation of high performance and vibration free concrete.

There also are smaller groups, who are working with different things. One group, with people from the contractors and the Western Region, are working in order to get more own suggests from the contractors in the tenders.

7.4 *International co-operation*

When working groups, for solving of technical problems, are established, they can have participants from more then one country. The advantage with this is that there will probably be more angles of approach, and there can be a more unprejudiced discussion. The cost and time for travel is about the same in Europe as it is inside Sweden.

It is possible to establish a network of people from different road administrators in Europe, for certain knowledge areas. These networks can be similar to networks inside an organisation with regions, see chapter 4.2.

Members of international co-operation committees, like PIARC and CEN, always ought to give written reports. These members also ought to be included in the work with development, in their special knowledge area, in their own country.

It is also important to organise a company in such a way that it is easy to take in experiences from abroad.

8. ORGANISATION

8.1 *Organisation of SNRA*

The road administration in Sweden of today is organised in a head quarter and regional administrations (regions).

The head quarter is divided in "two halves", one

part with three departments take all the decisions, and another part with four departments supports the decision departments and the regions. One of the supporting departments (which is the largest department) consists partly of the bridge experts. The head quarter also includes the own production, which is divided in a design, construction and ferry "companies".

The regions have departments for purchasing consultants and contractors in order to plan, design, construct, repair and maintain the road network inclusive the bridges. Each project has a responsible project leader, and the project leader can be responsible for many projects.

The region has also departments for economy, authority, traffic safety and environmental questions.

A third part of the region has different departments with specialists, who shall support the other departments, especially the project leaders, with competence in different knowledge areas.

8.2 *Future development*

In the future there will be changes more often in the organisation of a road administration. A great risk with this that you loose competence and knowledge about bridges. On the contrary, with new people, it is a possibility to change old habits and get a more creative environment.

I don't think that it is good to spread the people, who are working with bridges, into different departments in an administration. There ought to be enough with people to get possibility for an active dialogue and co-operation. There also ought to be a mixture between experienced people and relative new employed people.

Some important things to take care of when an organisation is changed are:

- Try to create a creative environment among the people who are working with bridge.
- Try to save all the old and important knowledge.
- Try to create a spirit of co-operation and understanding as well inside the administration as in connections with consultants, contractors and technical universities.
- Try to get as low borders between the different departments as possible.

8.3 *Internal organisation examples from SNRA Western Region*

In the Western Region of SNRA all people, who are working with bridge maintenance, belongs to the engineering department, except the people, who purchases and control bridge repair, who belongs to the maintenance department.

During the last years, in the engineering department, there has been work with improving the specialist's competence, collecting and spreading the knowledge, lowering the borders between the department and a lot of other things.

In this year, a group has been working in order to create a system for collecting, storing and spreading experiences in an effective way.

9. TECHNIQUE IN DIFFERENT COUNTRIES

Different countries use different technique. This can depend on climate, raw material, experiences, tradition, established technique, price level, economy and a lot of other factors. One example is Great Britain, who use external posttensioned reinforcement, and very often preplaced concrete elements, in bridge constructions, while Sweden use internal posttensioned reinforcement and very much concrete placed in situ.

These circumstances can make a discussion about bridge technique between different countries difficult. However, if the discussion starts with an explanation of the background to the used technique, it can be very valuable discussions with a lot of new angles of approach.

10. CONCLUSIONS

- There are a great need for cost-effective strategies for maintenance and repair of bridges. Furthermore there is a need for understanding of the deterioration and ageing mechanisms. The most important part in these processes and systems are the people, who are working with maintenance and repair of bridges.
- All these people, who are working with different kinds of bridge maintenance, must have competence and knowledge in many subjects. They must have experience and technical knowledge about theories and deterioration mechanisms, and they must also understand the background, why things happen. Other important qualities are their creativity, ability to sell good ideas, interest for new ideas, working without prestige and ability to co-operate.
- Knowledge is one of the most important parts of the competence. There is a lot of knowledge about bridges, in different organisations and also in different countries, knowledge about different problems with design and deterioration mechanisms for materials, which are used in bridges. If the people, who are working with bridges, use all this knowledge, it is possible to save a tremendous amount of money.

- A system for collecting and storing experiences can be a useful tool testing theories and spreading knowledge.
- One important meaning with training is to give knowledge to the people, but there are also other important things, which can be trained, in order to improve the competence. It can be creativity, ability to sell good ideas, interest for new ideas without prestige and ability to co-operate etc.
- Some trends in the society of today are probably pushed by the market economy. This has meant that the road administrations are working more as only clients, and the "own production" is working more and more as "companies. This trend is followed by the tendency towards quality assurance for the companies, functional demands on the products and calculations of whole life costing (WLC) and environmental impact (LCA) for the projects. The technical development and the development of the computer systems (BMS) are also going to influence the future.
- Tomorrow there will be many actors, who are working with, and have competence in, bridge maintenance. There are a lot of consultants for the design and many contractors for repair and construction. This makes it more difficult to spread the knowledge. Because of that, it will be very important to have a co-operation between different actors on this market. Co-operation is also a very good competence training for individual people, and co-operation is also a good way of spreading knowledge.
- In the future there will be changes more often in the organisation of a road administration. A great risk with this development is that you loose competence and knowledge about bridges. On the contrary, with new people, it is a possibility to change old habits and get a more creative environment.

Operation and maintenance assessment and requalification of structures and equipment

Operation and Maintenance of Large Infrastructure Projects, Vincentsen & Jensen (eds)
© 1998 Taylor & Francis, ISBN 90 5410 963 7

Strategies for monitoring and inspection of structures and installations

Clas-Göran Rydén
Swedish National Road Administration, Borlänge, Sweden

ABSTRACT: A strategy for monitoring and inspection of bridges is based upon bridge maintenance goals, which in turn are based on the political goals of road management. The strategy for monitoring and inspection of bridges requires different types of inspection, each containing a wide set of measurement methods. The different inspection types are described, as well as examples of the related inspection procedures and measurement methods. Some new measurement technologies are discussed. Impulse radar measurements for inspection of waterproofing on concrete bridge decks, is also used as an example for discussion of difficulties in implementation of new non-destructive measurement technologies.

1 INTRODUCTION

The political goals of road management can be divided into the following categories :
- Traffic safety
- Protection of the environment
- Transportation efficiency
- Cost-effectiveness in infrastructure management and maintenance.

Since these goals are to be reached under financial limitations, there is an obvious need for planning tools containing optimising procedures. The maintenance strategy will serve as a basis for developing the necessary planning tools.

For the case of bridge maintenance, these broad political goals can be condensed to the following bridge maintenance goals :
- Fatalities or severe injuries, caused by bridge failures, are not accepted.
- Bridge maintenance, like all other infrastructure related activities, shall be carried out without using material or procedures that is harmful to the environment.
- The life-cycle cost, including traffic related costs, shall be minimised.

2 STRATEGY FOR MONITORING AND INSPECTION

For the case of bridge inspection, the maintenance goals leads to the following basis for strategy for monitoring and inspection :
- Deficiencies, which may be dangerous to the public, shall be revealed, and proper action should be taken immediately. For the case that such deficiencies are not going to be obvious, constant monitoring of the critical component may be necessary.
- Deficiencies and damages shall be revealed as soon as possible, giving time for planning (and procurement) of maintenance measures, or repair work.

A strategy for monitoring and inspection must also take a number of other factors in consideration. Some important factors are
- The competence of inspection staff
- Equipment for inspection and measurement
- The cost (including cost of traffic disruption) of carrying out inspections.
- Time between inspections
- The acceptable time for a deficiency to exist.
- The rate at which a deficiency develops into a major damage.

– The degree of development in which a damage is possible to detect
– Deficiencies and major damages, that constitutes a threat to public safety.
– The cost (including cost of traffic disruption) of fixing a minor deficiency.
– The cost (including cost of traffic disruption) of major repair work.
– The intended service life for components and structures.

This leads to a strategy for monitoring and inspection, that requires several inspection types, ranging from the regular inspection that
– is fast and cheap,
– is carried out in short time intervals,
– requires only little bridge engineering competence,
to the Special inspection that
– Is expensive and sometimes time-consuming,
– is focused on one type of problem, or one type of bridge component,
– requires special competence.

3 TYPES OF INSPECTION

The a strategy for monitoring and inspection requires different types of inspection, each containing a wide set of measurement methods. The types of inspection are as follows: Regular inspection, Superficial inspection, General inspection, Major inspection and Special inspection.

3.1 Regular inspection

The inspection shall have the aim of detecting acute damage which may affect the safety of traffic and the integrity of the structure in the short term.

The inspection refers to the top of the bridge and to the road embankments on each side of the bridge.

The inspections are to be made regularly by the maintenance contractor; it is appropriate for this to be done in conjunction with the inspection of the road network.

3.2 Superficial inspection

The inspection shall have the aim of verifying that the requirements specified in the maintenance contracts are complied with.

The inspection refers to those structural members and elements for which certain requirements have been specified as to their properties and the action to be taken.

The inspections are to be made by the maintenance contractor at least twice a year for bridges on the national road network and at least once a year for other bridges.

The inspections are to be made by personnel who have good knowledge of the appropriate methods of measurement and are familiar with the structural design and mode of action of the bridge.

3.3 General inspection

The inspection shall have the aim of following up the assessments, made at the time of the immediately preceding major inspection, regarding damage which has not been put right. The inspection shall also have the aim of detecting and assessing damage which would have resulted in unsatisfactory bearing capacity or traffic safety or would have given rise to substantially increased administration costs if the damage had not been detected before the next major inspection. The inspection shall further have the aim of checking that the requirements specified in the maintenance contracts have been complied with. Any deviations shall be measured up.

All structural elements except those below water level are to be inspected. Visual inspection methods may be applied. Parts adjoining the bridge such as road embankments, slopes, abutment ends, fill, revetment and fenders are also to be inspected. The inspection is to be carried out at such a distance, with the assistance of any optical aids which may be necessary, that the above aims are achieved. If deviations are found from the assessments made at the time of the immediately preceding major inspection regarding damage which had not been put right, or if new damage is detected (as above), such damage is to be assessed from a distance within arm's reach in accordance with the requirements applicable for major inspection.

The inspections shall be made at intervals of not more than three years this includes major inspection. This requirement applies only to bridges where the theoretical span of the longest span exceeds 5.0 m. For other bridges a general inspection is to be made when necessary.

The inspections shall be made by personnel who comply with the requirements specified for major inspection.

3.4 Major inspection

The inspection shall have the aim of detecting and assessing defects which may affect the function of the structure or traffic safety within a ten year period. It is also the intention to detect defects which, if not remedied within this period, may give rise to increased administration costs. The inspection shall also aim to check that the requirements specified in the maintenance contracts are complied with. Any deviations found shall be measured up.

All structural elements (including those below water level) are to be inspected. Visual inspection methods may be applied. Parts adjoining the bridge such as road embankments, slopes, abutment ends, fill, revetment and fenders are also to be inspected. The inspection also covers mechanical and electrical equipment in a movable bridge. At the time of the major inspection, the required measurements are also to be made to determine, inter alia, bottom profiles chloride content and carbonation of concrete reinforcement corrosion cracking in steel structures. The inspection is to be carried out from a distance within arm's reach.

Inspections shall be made at intervals not greater than six years. The first major inspection of a new bridge is to be made just before the guarantee inspection, but not later than six years after the bridge was opened to traffic.

The inspections are to be made by personnel who possess the following competence:
− engineering training
− training as an inspector by the Swedish National Road Administration
− knowledge of the durability of bridge structures and the degradation processes to which these structures are exposed
− knowledge and experience required to predict development of damage
− knowledge and experience required to find appropriate technical and economical solutions to remedy damage
− knowledge of Regulations for Bridges BRO 94 of the Swedish National Road Administration, Regulations for Concrete Structures BBK and Regulations for Steel Structures BSK.

For underwater inspection it is necessary in addition to the above for the personnel concerned to have the required certificate for work under water. For inspection of mechanical and electrical equipment the following applies in addition to the above:
− Electrical competence in accordance with the Electrical Installations Ordinance
− knowledge required to perform trial runs on machinery and electrical equipment
− knowledge and experience of hydraulic equipment
− knowledge of the Swedish Regulations for the Design and Maintenance of Electrical Installations

3.5 Special inspection

The inspection is to be carried out as and when necessary to investigate in greater detail the defects which were detected or were presumed to exist at the time of the regular inspections. An example of such investigation is pulse radar measurement for inspection of waterproofing on bridge decks. Inspections are also made on the mechanical and electrical equipment which actuates the opening mechanism of movable bridges.

Mechanical and electrical equipment on movable bridges shall be made at intervals not greater than three years, this includes major inspection. Butt welds in primary loadbearing elements are inspected in conjunction with the major inspection made prior to the guarantee inspection, but not later than six years after the bridge had been opened to traffic. The times of inspection of other details are determined when the regular inspections are made.

For inspection of mechanical and electrical equipment, the requirements set out for the major inspection apply. For most other inspections also, the competence requirements specified under the major inspection apply. Specialist competence is required for instrument measurements such as Ultrasonics, Radiography and Thermography

4 MAJOR INSPECTION OF ELECTRICAL EQUIPMENT AND INSTALLATIONS

The inspection is to be carried out in two stages, visual observations and trial runs.

4.1 Visual observations

All electric components shall be visually observed for signs of malfunction, wear or security hazards. Examples of visual observations are given in table 1 below .

Table 1. Examples of visual observations of electrical equipment and installations

Example of Unit or component	Example of check to be made
Hydraulic unit	– that the position indicators or LED on solenoid valves work, that the detectors for level, pressure and heat work, that there are no oil leaks on electrical equipment or cables.
Motors	– that slip rings/commutators are not burned and that the brushes are in uniform contact. The space shall be free from dust from the brushes. There must be no sparking during operation – that the brushes are not worn more than is commensurate with full contact along the whole brush holder. Insulation tests should be performed if the motors are above a certain age. The insulation level shall be better than1.4 Mohm. – that the insulation on the windings is not damaged.
Resistors	– Check that the resistor chamber or coils are not damaged – that there is no overheating. Is there discoloration on resistor elements or cover plates? – that resistor fingers are not deformed. Are there grooves due to partial shortcircuiting or have they become welded together? – that mica insulation on fingers has not been displaced. Porcelain insulators and tubes shall be undamaged
Connections	– Check that connections are tight. See if there are any traces due to overheating, discoloration of connections due to over-heating and/or material changes (alteration of flexibility).
Brakes	– Brake retractor/brake thrust – that the cable to the brake is flexible and whether there is any movement in the brake frame

4.2 Trial runs

Trial runs should be made when traffic is of low intensity.
Trial runs shall be made using all conceivable operational alternatives:
– Operation from the control desk
– With the standby plant
– Different motor alternatives
– Redundant drives
– Different speeds
– Emergency machinery
– Remote control desk

4.3 The following functions should be checked during trial runs

– Manual operation interlocks
– Interlock between different items of
– machinery and signals.
– Timing circuits
– Delay between red light and barrier
– lowering > 10 s. Delay in centering
– swing bridges etc.
– Override facilities, barrier interlocks
– etc.
– Battery backup for barrier lights.
– Emergency stop and normal stop in intermediate position. Avoid a stop where there is a risk of damage.
– Start in intermediate position
– Interlocks between places of operation.
– Connection of sea signals.
– Operating times
– All indications during operation. Current indications for signals must not fail.
– Instruments. Note currents at start and during operation of machinery.

5 MAJOR INSPECTION OF MECHANICAL EQUIPMENT

The inspection is to be carried out in two stages, visual observations and trial runs. Trial runs of mechanical equipment are carried out in the same manner as trial runs of electrical equipment.

5.1 Visual observations of mechanical equipment

All mechanical equipment shall be visually examined for signs of wear or other types of deterioration. Examples of equipment and their corresponding observations are given below.

Check gear drives. Does the gearing engage correctly and is wear normal? Are there incipient cracks at the roots of the teeth? Enclosed gearing should be examined carefully for leakage of lubricant. This applies in particular to small gear assemblies which are emptied rapidly without any external signs. Gearing always generates a certain amount of friction heat. When a gear is getting worn, temperature rises. Gearing temperature also increases due to excess lubrication and overload. The temperature of gearing should therefore be placed under observation. Any change in gearing usually also causes a change in gear noise. Measurements of sound level can then be made. Defective gearing can also give rise to a different vibration pattern. Observations with vibration measurements can then be made.

Check that shafts are straight and that shafts at the fillets or otherwise are free from surface cracks. Are shafts correctly mounted, and are shafts in an interconnected system parallel?

Check bearings. When a sliding bearing is getting worn, bearing play increases, and this is in many cases a good indication of the condition of the bearing. Bearing play can be measured in different ways, for instance with a dial gauge, a lead impression in combination with a micrometer, or with a feeler gauge.

Check chains and chain wheels. Wear at the pins is usually the best indication of the condition of a roller chain. One simple and reliable way of noting the wear at the pins is to measure the length of the stretched out chain over e.g. ten links and to compare this with previous measurements. Make sure that the tension is correct.

Check couplings. Shaft couplings are important elements which are found on many bridges. Couplings which are worn or loose can cause serious problems, and they should therefore be checked carefully. There are different types of couplings. The most common type is the flexible rubber disc coupling in which rubber discs are the damping elements. The rubber discs are worn radially and should be replaced when the play exceeds 1 mm. The coupling continues to work when the discs are worn, but the impact forces which are generated can damage the connected elements. Another type of coupling which is often used is the claw coupling which, in the axially displaceable version, is used for manual operation. When the play is greater than 1.5 mm, the coupling should be replaced.

Check brakes. Slip is a usual fault. The cause may be wear of the brake lining or the presence of some deposit such as oil on the drum. The time for the brake to act is then longer than normal. Slip may also be due to excessive play between the drum and the lining, or wear and misalignment of the linkage system.

6 MEASUREMENT METHODS FOR MECHANICAL EQUIPMENT

Examples of current measurement methods is given below.

6.1 Crack depth

If an electric current is impressed on the surface of a piece of metal and the voltage drop is measured over a certain distance, the voltage drop on a surface free from defects is the same regardless of where the measurement is made. If there is a crack in the region where the measurement is made, the voltage drop increases. The deeper the crack, the larger is the voltage drop.

6.2 Vibration measurement

A machine which has abnormally high vibration wears out more rapidly. Even an imperceptible rise in vibration level increases the temperature, and the bearings are worn more rapidly. Special, portable or stationary, equipment is available for vibration measurement.

6.3 Shock pulse measurement

The shock pulse method (SPM method) measures the magnitude of the mechanical shocks generated in damaged rolling bearings. Different makes of meter are available.

6.4 Spectrometric oil analysis

Oil samples are taken at definite intervals in a special sampling bottle. The quantity of every element is measured by an atomic absorption spectrophotometer in p.p.m., i.e. the number of parts of the element per million parts of oil. The method is used for oil where moving parts are in contact.

7 MAJOR INSPECTION OF LOADBEARING ELEMENTS OF STEEL

Where this is considered necessary, bridges with a steel superstructure should be observed both with and with-out traffic. This is appropriate, for instance, when large oscillations occur in the bridge or there are abnormal movements at the supports.

7.1 Corrosion

Check the surface finish on steel structures for flaking, blistering and corrosion. Impurities, sand, bird droppings etc. can bind moisture and greatly accelerate corrosion. Sensitive details are bolted and riveted connections, supports and places where girders are in contact with the concrete structure, and on a steel bowstring bridge the junction between the arch rib and the tie.

A check is to be made inside steel box girders whether there is any condensation which increases the risk of corrosion.

Check whether there is any crevice corrosion in bolted and riveted connections. If crevice corrosion is suspected, the thickness is to be checked by ultrasonics so that the reduction in thickness due to rusting can be assessed.

Special inspection is to be made of steel culverts at the water line where corrosion is most common. It is best to carry out the inspection at low water. Where there is pitting, the remaining sheet thickness is to be determined, either by taking a sample or by ultrasonics. When corrosion has penetrated right through a steel culvert, there is a risk that the fill around the culvert will be washed away. This means that the structure loses its loadbearing capacity.

7.2 Cracking

Check steel structures for cracking and along the web over the entire length of the bridge and on both sides. A special inspection is to be made if necessary. Check that no welding or cutting has been performed on older rolled girders. If the girders were made before 1952, the material is basic Bessemer steel. In this steel, cracks or fracture easily occur when it is welded or cut. Any crack found must be immediately treated. The first action to be taken is to stop the crack by drilling a hole at its end.

7.3 Loosening

Check that no bolts or rivets are missing or sheared off. Nor shall bolts or rivets be loose or ineffective. A loose bolt/ rivet can be detected most easily by checking if the paint around it has cracked.

7.4 Movement and deformation

If movement (slip) is found in high strength friction grip connections, check the preloading of the bolts.

Check whether suspenders are abnormally tight or loose by comparing their 'ring' in response to a light tap. One good way of checking is to watch the bridge under heavy traffic and see whether any suspender oscillates too much. A check is also to be made whether the suspender attachment in a suspension bridge has moved along the main cable. Suspender attachments may also be deformed due to impact by snow clearing equipment.

Check flanges, web plates, tie bars and struts, suspenders and ties for any kind of damage, buckling or deflection.

Check also for abnormal deflection of the structure. This can be studied most easily by following the alignment of the parapet.

8 MEASUREMENT METHODS FOR STEEL STRUCTURES AND COMPONENTS

Measurement on steel structures are carried out as non-destructive testing, as well as sampling and destructive testing of samples. Measurements are made in order to detect internal and external defects in welds and parent material, and to investigate the properties of the steel.

Surface treatments on steel is investigated by means of non-destructive testing methods such as coating thickness measurement, and electric void detection. Adhesion of paint on steel is measured by destructive pull-off test.

8.1 Problems

Problems in older structural steel are often associated with uneven quality and inadequate ductility. The basic Bessemer process often produced steel with high contents of phosphorus and nitrogen. This had the result that the material was prone to ageing and increasingly brittle as time went on. Brittle fracture is characterised by the suddenness of failure and very little plastic deformation. It is an unstable

failure, i.e. the energy stored in the structure is sufficient to maintain the failure process. The most common causes of brittle failure are listed below in rank order.
– Steel with high contents of impurities
– Low temperature
– Mechanical damage such as dent made by an impact.
– High loads
– Welds in old steel

It is a general rule that older steels shall not be welded unless the material has been analysed and the other strength properties have been established. Normally, however, additions, alterations and repairs are made with bolted connections. If welding is to be carried out on older steel structures, an investigation must first be made whether this is realistic. This means that a weldability investigation must be made.

8.2 Non-destructive testing of steel

Non-destructive testing on steel can be made by the magnetic particle method and the liquid penetrant method. Defects or imperfections which are detected are normally followed up by radiography or ultrasonics.

8.3 The magnetic particle method

The magnetic particle method is used to indicate cracks, folds, pores, undercuts and similar defects on or very near the surface. Magnetic particle tests are to be made in accordance with SS 11 44 01. The higher the standard of surface finish, the greater the sensitivity of the method. The method can only be used on ferromagnetic materials. Paint and impurities must be removed before the test. In order to facilitate visual examination, paint of a contrasting colour (white strippable lacquer) is to be applied before the test.

Both a permanent magnet and an electromagnet can be used in the test. The area to be tested is magnetised with the magnet. During magnetisation magnetic powder made into a slurry with a suitable liquid such as paraffin is applied to the workpiece. If a crack is found, the flow of magnetic powder is diverted and a leak-age flow occurs across the crack. This leakage flow causes magnetic polarisation of the edges of the crack. It is convenient for defect indications to be recorded by photography.

8.4 Liquid penetrant testing

Liquid penetrant testing is used to indicate cracks, folds, pores, undercuts and similar defects on or near the surface. The higher the standard of surface finish, the greater the sensitivity of the method. Liquid penetrant testing is to be carried out in accordance with SS 11 45 01. The method can also be used for non-ferromagnetic materials such as aluminium. Prior to the test, paint, oil, water, dirt and other impurities must be removed from the intended work area. A penetrant liquid is sprayed, poured or painted on the surface. After some time (5 seconds - 30 minutes) the liquid will have penetrated into any cracks or other surface defects. Excess penetrant liquid is then removed with a cleaning fluid, and a white liquid called developer is applied.

If there are defects, the penetrant is drawn to the surface where it spreads into the developer. After developing the surface is examined in a suitable light. It is convenient for defect indications to be recorded by photography.

8.5 Radiography and Ultrasonics

Welds are inspected to detect internal and external defects. Radiographic and ultrasonic tests are methods in common use to check for, and detect, internal defects in welds. Examples of defects are porosity, non-metallic inclusions, root defects, incomplete penetration and internal cracks. Radiographic and ultrasonic tests complement one another For certain types of cracking and incomplete penetration, ultrasonic tests are preferred.

Radiography can be applied only to butt welds. It is to be performed in accordance with Swedish Standard SS 11 41 01, Class B, with the image quality in accordance with SS 11 41 30, Class B. Radiograms shall be interpreted in accordance with SS 11 41 01 and SS 06 61 01.

Radiography is performed using equipment comprising an X-ray tube, cables and control panel. The X-ray tube generates a beam of rays of different wavelengths. The electrons in the beam are accelerated by very high voltage and are retarded by a suitably placed metal plate. Radiography is based on utilising the ability of the X-rays to blacken an X-ray film. Differences in intensity in the exiting radiation due to defects in a weld, for instance a crack, can be made visible by the differences in density on the film. An X-ray film consists of a transparent and flexible substrate coated on both sides with a photosensitive material. A radiogram is excellent documentation of e.g. welds, and can be archived for an unlimited length of time.

Ultrasonic tests can be used on all kinds of welds. They are to be performed in accordance with SS 11 42 01 and SS 06 61 01.

Ultrasonic tests are performed using ultrasonic test equipment. This comprises a transducer with both transmitting and receiving function, and an oscilloscope on which the results are presented. Ultrasonics sound waves can be propagated in solid bodies such as steel, but not in air.

The most common method is to apply the ultrasonic beam to the steel material with a transducer. The sound is transmitted into the material via a couplant which is a thin film of oil or water.

8.6 Destructive testing of steel

If it is necessary to investigate the strength of the material, sampling and testing should be performed in accordance with Swedish Standards for acceptance tests on metallic materials.
- Tensile tests in accordance with SS 11 01 20 and SS 11 21 10.
- Impact tests in accordance with SS 11 01 51, SS 11 23 51 and SS 11 23 52.
- Bending tests in accordance with SS 11 01 80 and SS 11 26 26.
- Determination of hardness in accordance with SS 11 25 10 HB, SS 11 25 16 HV and SS 11 25 12 HRc.
- Chemical analysis in accordance with SS 11 01 05.

8.7 Bolted and riveted connections

Checks on connections with normal tightening are made with a large spanner. Loose nuts are tightened and secured against loosening by a punch mark on the threads.

Checks on high strength friction grip connections with regard to the specified tightening torque for the bolts are made with an accurately calibrated torque spanner. At the time of checking, the nut is to be tightened a little (ca 5°).

All rivets are to be checked by a light blow on the rivet head with a hammer (weight ca 0.35 kg). There shall be no movement in the rivet. This can be checked, for instance, by holding a finger against the edge of the rivet as it is struck.

9 MAJOR INSPECTION OF WATERPROOFING ON CONCRETE BRIDGE DECKS

The first indication of leakage through the waterproofing is cracking on the soffit of the deck slab and, if the bridge has water vapour outlets, discharge of water from these. After a short time efflorescence occurs around the leak. In the case of efflorescence near cracks and water vapour outlets, the presence of a leak cannot be determined with certainty unless the soffit of the slab is damp around these, since efflorescence may have occurred before the waterproofing was laid.

A damp patch on the soffit of the deck slab, where there is no cracking or a water vapour outlet, is not usually detected until long after leakage had started. Leaks of this kind can be detected by the presence of damp patches with or without efflorescence, differences in colour, rust or frost splitting and crazing, or a combination of these.

Pay special attention to prestressed concrete structures and to bridges with waterproofing laid several years ago. The same applies to bridges with older types of waterproofing such as coats of asphalt or epoxy. If there is crazing or cracking in the surfacing, in most cases this indicates weathering of the concrete protection course or damaged/blocked drainage channels. In the long run, these may give rise to damage in the waterproofing. Blistering on the surfacing may indicate that there is blistering in the waterproofing also. If it is suspected that there is leakage or damage to the concrete protection course, a special inspection of the waterproofing must be made.

10 SAMPLING AND MEASUREMENT OF CONCRETE SURFACES BELOW WATERPROOFING

When leakage is detected or suspected, a sample must in the first place be taken within the area assumed to be affected by the leakage. A reference sample should always be taken from a surface where no leakage is suspected.

Samples are conveniently taken in "windows" by drilling a core through the surfacing, concrete protection course if any, and waterproofing. Drilling is to be carried out in stages, course by course. Each course is to be inspected before the next level is exposed. On a bridge with gravel carriageway or gravel surfacing, the courses above the waterproofing and concrete protection course must be removed. This can be done by digging a pit, or if the bridge has surfacing thicker than 0.5 m, by digging inside a precast concrete ring.

10.1 Investigation of concrete

The investigations of the concrete in the deck slab which may have to be performed are as follows:
– Visual checking of concrete surfaces inside the "windows"
– Measurement of chloride content
– Tests for frost resistance
– Tests for strength
– Visual checking of drilled samples
During visual checks in windows, an assessment is made whether the concrete is weathered or eroded. The depth to which the damage has penetrated must also be studied. If the damage extends down to or below the top reinforcement in the slab, the extent of damage must be established. The concrete is chipped away with a hammer or cold chisel. The fracture surfaces are carefully studied. They can provide useful information regarding the state of the concrete.

11 SOME SAMPLING AND MEASUREMENT METHODS FOR CONCRETE BRIDGE DECKS

Sampling on concrete structures is mainly carried out to determine compressive and splitting strength, frost resistance, chloride ion concentration, degree of carbonation and thickness of concrete cover.

In determining the strength of a bridge, an overall view is essential. Both the "best" and "worst" positions are therefore of interest for the test. If only the "best" positions are chosen, the load-bearing capacity of the bridge can be overestimated.

11.1 Cores

Cores are drilled out of the completed structure. The term test specimen refers to that part of the core which is used in laboratory tests. When a core is drilled out of a concrete structure, a diameter of 100 mm is to be aimed for. The length of cores shall be such that the intended tests can be carried out. For compressive and splitting strength, a test specimen length of 100 mm is required for each test. With a test specimen of this dimension (100 * 100 mm), the results can be directly compared with those for a cube of 150 mm sides. The term test specimen in this context refers to a part of the core which is not

affected by e.g. reinforcement, decomposed concrete, etc.

When cores are taken, the sampling sites and the cores must be documented on a sketch, and the cores also marked in relation to this. The sketch shall accompany the cores to the testing institute. Swedish Standard SS 13 11 13 specifies how cores shall be drilled and treated for strength determination.

Apart from drilling out cores, in situ tests for determination of strength can be carried out by various other methods such as:
– Ultrasonic measurement
– rebound hammer test
– combined ultrasonic and rebound hammer test
– indentation test with a ball hammer
– penetration test, for instance the Windsor probe test
– core bending test, e.g. the TNS method or the BO test

If compressive strength is determined in a way other than by using drilled cores, less reliable results are obtained. One of the reasons is that the compressive strength is measured indirectly via other parameters. However, in structural members which are sensitive to having large bits of concrete drilled out, the above methods of measurement can provide a useful supplement. Determination of strength in the completed structure using e.g. the above methods requires a special permit.

11.2 Depth of carbonation

Carbon dioxide which penetrates into concrete from the air reacts with the calcium hydroxide in concrete. This process, called carbonation, results in a lowering of the pH value of concrete from 12-13 to 7-8 and thus in a considerable reduction in the protection provided against the corrosion of reinforcement. This lowering of the pH value can be shown by a change in the colour of a suitable indicator. The indicator liquid comprises 3.0% by weight of phenolphthalein in ethanol (denatured). This liquid turns red in an alkali environment (pH>9.2) but remains unaltered at lower pH values.

Determination of depth of carbonation is described in SS 13 72 42. Tests may also be made directly in the completed structure by chiselling a hole for testing. The indicator liquid is sprayed in after the hole had been blown clean. The test shall be made on a surface which had not previously been in direct contact with air. The concrete surface is carbonated relatively rapidly, and an erroneous result can therefore be obtained.

11.3 Concrete cover

Electromagnetic measurement of the concrete cover is based on the interference which the reinforcement causes with the magnetic field generated by the instrument. With some instruments, both the cover and the bar diameter can be estimated. The depth of penetration varies, but in some instruments it may be as much as 100 mm. The diameter of the reinforcing bar must generally be known. Where there are several layers of reinforcement, interpretation of the readings may be difficult.

In order to calibrate results from electromagnetic measurement methods, the reinforcement shall be exposed by breaking out some of the cover.

11.4 Electric potential measurement

Rust stains on the concrete surface and delamination of the concrete cover are obvious signs of corrosion of reinforcement in concrete. At that stage, the corrosion damages have developed to a degree at which extensive concrete repair work is needed. Ongoing corrosion of reinforcement must therefore be detected long before it has had time to assume harmful proportions.

Whether or not conditions exist for reinforcement corrosion is assessed by measurement of electric potential. Such measurement may be performed in accordance with ASTM C 876. In order that it may be used as a decision base, the method must at all times be complemented by measurement of the electrical resistance of the concrete.

Measurement of electric potential alone will not give an answer to what degree the reinforcement has corroded. Measurement of the rate of corrosion is possible, but since the actual rate of corrosion usually fluctuates within a wide range, accumulated corrosion damage is still difficult to predict. For this reason, reinforcement must be exposed and visually examined at some points.

Measurement of electric potential is not a suitable method for locating corrosion in concrete under water. Considerable corrosion may occur without the generation of corrosion products which cause swelling. The only reliable method which is available today for the detection of reinforcement corrosion under water is to visually examine a cleaned concrete surface. This may need the assistance of a diver.

11.5 Compressive and tensile strength

The strength of the test specimens taken from the cores shall be determined in accordance with Swedish Standards. The shape, dimensions and tolerances of the specimens are set out in Swedish standard SS 13 11 11. The compressive strength of the specimens is determined in accordance with SS 13 72 30, and the tensile (splitting) strength in accordance with SS 13 72 13.

The mean and standard deviation are calculated for all test specimens and are normally quoted in the test certificate together with the individual values.

11.6 Chloride content

The concentration of water soluble chloride ions as a proportion of the weight of cement is determined at different levels in a concrete structure. On the basis of these values, a chloride profile can be plotted which shows the chloride ion concentration as a function of the distance from the surface of the concrete. The chloride ion concentration can be determined in accordance with SS 13 72 35. Tests according to this standard are performed only in well equipped laboratories. Usually, however, the chloride ion concentration can be determined in both the field and the laboratory in accordance with the RCT method.

11.7 Frost resistance

The term frost resistance denotes the ability of concrete to resist repeated freezing. Frost resistance is dependent on air content and on the size and distribution of the air bubbles. Frost resistance is enhanced if there are many small air bubbles, water-cement ratio is limited (<0.45) and vibration and curing are performed with care.

For the investigation of frost resistance, test specimens of 100 mm diameter should be used. The laboratory investigation is performed on a 50 mm thick test specimen sawn out of a core, which is wetted in a bath of chloride solution or pure water. The specimen with its bath is subjected to repeated freezing cycles. One cycle usually takes 24 hours to complete.

12 REQUIREMENTS FOR FUTURE INSPECTION METHODS

The road traffic of tomorrow will put new demands on development of new inspection methods and

procedures. Future inspection methods will have the following characteristics :
– Not disrupting traffic. The cost of traffic disruption often is the single most important portion of the total inspection costs.
– No working hazards. In practice, this will mean that the place of work must be separated from the traffic.
– Quick, cheap and efficient. Modern technology has improved the possibilities for efficient data acquisition and storage, as well as automatic positioning of measured data.
– The quality of measured and position data must be described. This enables the use of data for other purposes that were not foreseen at the time of collection.
– Ability to detect and assess deterioration well in time before drastic repair work is needed. This allows early low-cost maintenance measures as well as plenty of time for planning and optimisation of maintenance work.

13 IMPULSE RADAR FOR BRIDGE DECK INSPECTION

Before the surfacing and waterproofing are opened up for inspection of the concrete situated below, impulse radar measurement of the waterproofing should be made.

13.1 Background

During the last 15 years, there have been substantial problems with damages on concrete bridge decks caused by concrete deterioration and reinforcement corrosion. The main reason is leakage of the waterproofing membrane that is supposed to protect the concrete from salt and water, see figure below.

In Sweden, approx. 7 000 bridges, out of 12 000, have bridge decks made of concrete that is not salt-frost resistant, i.e. the concrete will deteriorate quickly when subjected to freezing and thawing in the presence of salt.

Since it has been known for a long time, that both moisture and presence of salt affects the dielectric constant and impedance of concrete, a non-destructive bridge deck measurement method have been developed, based on impulse radar (also known as Ground Penetrating Radar), and computer-based interpretation of radar signals. This method have been in use for extensive trials in 1993 to 1995.

13.2 General Principles

A radar pulse is transmitted from a radar antenna. Seen on an oscilloscope, the pulse will appear as a sharp peak, which means that the pulse contains a wide range of frequencies. As in other cases of wave transmission, interfaces between different materials will create an upward-reflected signal, as the radar pulse moves on downward. This reflected signal is received, usually by the transmitting antenna, recorded and stored for further processing.

13.3 Basic requirements

The basic requirements for the measurement method is the ability to locate areas of bridge deck having abnormal microwave reflectivity.

For practical reasons, measurements must be made from a vehicle with a linear array antenna mounted transversely to the travelling direction. It is desirable to cover the largest possible width in one operation. This makes matrix operations of signal data more accurate. The data collection process must also be fast enough to allow a high travelling speed during measuring, for productivity reasons. In order

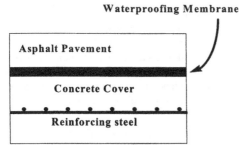

Figure 1. Typical cross-section through upper part of a concrete bridge deck

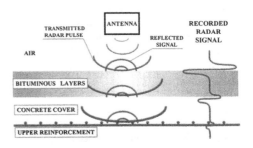

Figure 2. Transmitted pulse and reflected radar signal

not to disturb the traffic, the array antenna is designed to cover one lane width (2,5 - 3 m), and the data collection system allows measurement to be carried out at normal traffic speed.

13.4 Interpretation of data

What we are looking for in this case are areas where the reflected radar signal from the waterproofing/concrete interface looks different from the "normal" case. Such anomalies can be caused by a number of phenomena see examples in table 2 below :

Table 2. Phenomena affecting radar reflex amplitude from waterproofing/concrete interface

Increases Reflex Amplitude	Decreases Reflex Amplitude
Dry porous pavement	Wet road surface
Ponding on top of waterproofing	Salt on road surface
Moist and salt in concrete	Porous concrete

Not all of these phenomena are of interest, in fact measurements are preferred to be carried out in the summer after long periods without rain, so that there will be no salt or water neither on the road surface, nor on top of the waterproofing membrane.

13.5 Field trials

The equipment used consisted of a vehicle-mounted array antenna having 5 units with 500 MHz centre frequency, covering a lane width of 3,0 m. The antenna units sent 5 channels of parallel data into data storage system. A sufficient number of measurements were made on each bridge, in order to cover the entire bridge deck width.

Evaluation of data, which were done after measurements, began with choosing a "time-window" in order to extract the signals from the concrete/waterproofing interface from other signals. Data from parallel "measurement lanes" were then rearranged in a co-ordinate system describing the bridge deck area.

Results of radar measurements are presented in three different forms .
– A colour-coded rectangular surface map, representing the bridge deck, showing relative reflex amplitude from the waterproofing/concrete interface, see figure 3.

– A rectangular surface map, representing the same bridge deck, showing recommended sampling points, see figure 4.
– A table of co-ordinates for recommended sampling points of the bridge deck, see table 3.

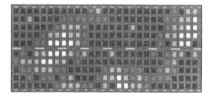
Figure 3. Colour map of the bridge deck

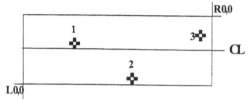
Figure 4. Recommended sampling points on the bridge deck.

Table 3. Example of co-ordinates for recommended sampling points of the bridge deck in figure 4.

Point no	L / R	Distance from edge beam (m)	Distance along bridge (m)
1	R	4,4	88,6
2	L	0,9	67,2
3	R	2,3	8,2

During 1993, 70 bridge decks were measured with this method. The bridges have been chosen in order to represent a typical range of bridge ages and pavement systems.

Core samples were taken in recommended sampling points, usually 3 cores per bridge. The equal number of cores per bridge deck were also taken at random locations, representing the normal condition of that bridge deck. The core samples were examined in laboratory, with respect to structural properties, as well as dielectric properties.

13.6 Problems experienced with impulse radar measurements

During field trials, a number of problems have been

identified. In order to increase the future efficiency of this method, these problems will have to be solved:
- Bridge decks are seldom rectangular areas.
- Weather conditions are not always perfect. Long warm and dry periods are rare in northern Europe. The voids in the pavement will always contain water to some degree.
- We are interested primarily in concrete degradation, and not pavement porosity variations. As roller compaction energy is restricted when laying pavement on bridges, the pavement on a bridge deck will most likely be more porous, than the pavement of a road. Moreover, the porosity will also show greater variations on the bridge deck.
- The actual thickness of the pavement and waterproofing can be calculated from drawings, but such calculations are not reliable.

14 OTHER NON-DESTRUCTIVE MEASUREMENT METHODS FOR EVALUATION

New measurement methods are evaluated as they become accessible. If the initial evaluation shows possibilities for future use, field tests are carried out.

14.1 Radiography of concrete structures

Radiography and gamma radiography can be used to locate reinforcement or other embedded fixtures. Cavities can also be detected. The equipment is suitable for detailed studies of faulty areas which have been located in some other way. In larger concrete structures such strong radiation sources are required that comprehensive safety arrangements are necessary. Sometimes, the safety arrangements will have to be so extensive, that the investigation can not be carried out as intended.

14.2 Acoustic measurements in concrete

Ultrasonic measurement is a common method used to check for, and detect, internal defects in e.g. metals. At present it is not possible to detect delamination or cavities in concrete. Development of automatic testing methods based on acoustic and ultrasonic techniques is in progress.

A measurement probe has been developed and the acoustic scanning technique has been proven to successfully image a delamination and to map areas of porous concrete. The extent of damage in the scanned plane can be quantified through measurement of acceleration only. No quantitative information on the flexibility of the structure can however be extracted. Higher frequency components which could be associated with propagating waves have not been identified with the present equipment. At present determination of the location (depth) of a defect can not be made with the equipment, though other non-scanning acoustic methods have shown it to be possible.

The results from scanning measurements using ultrasonics show that the technique needs to be further developed in order image defects. The results do not show images clear enough for conclusions to be drawn on the cause of registered patterns.

Registering simultaneously force and acceleration can be used in future acoustic systems to provide quantitative information on the three dimensional extent of damaged volumes.

14.3 Thermography

Thermography is a method in which the surface of interest is photographed with an infrared camera. Under favourable conditions, cavities and delaminations can be detected. It is obviously necessary for these defects to give rise to temperature differences. In northern Europe, sunlight intensity is not high enough to produce discernible temperature differences from delaminations in the structural concrete of the bridge deck.

Thermography has, however, been successfully utilised as a quality assessment tool for placing of waterproofing on concrete bridge decks. Even very small air voids becomes clearly visible.

14.4 Modal analysis of concrete bridges

Modal analysis of concrete structures, for instance bridges, is a method of evaluation which is under development. Briefly, the bridge is made to oscillate and resonant frequencies, mode shape, damping etc. can be measured. By monitoring a structure during several measurement events separated in time, an estimate can be made of the development of any damage. Modal analysis is principally useful in determining whether there are losses of prestress or fatigue fractures. In Sweden, this method has been used on a trial basis for prestressed and mild steel reinforced concrete bridges.

The interpretation of modal analysis data, usually requires comparison between modelled and measured dynamic behaviour, where the differences are interpreted as imperfections in the actual structure. Some common types of bridges, such as the conventional slab-frame- type concrete bridge, are however, difficult to model.

15 CONCLUSIONS

15.1 Acceptance of impulse radar measurements

It has proven difficult to implement this measurement method. There are a number of reasons for this, some of these may be generally valid for implementation of new technology.

At the present stage the following problem can be solved: "We suspect that the waterproofing membrane is not functioning properly. We don't want to cause extensive damage to the waterproofing membrane, in case it should be OK. Where should we take drilling cores for visual inspection ? ". It must be emphasised that impulse radar measurements does not tell if the waterproofing is damaged or not.

This means that the responsible maintenance engineer will still have to make critical decisions based upon experience and judgement. Although impulse radar measurements can be proven to be economically profitable, the new measurement technology does not, at the present level of development, appear to improve the basis for decision-making.

15.2 The cost of impulse radar measurements

The cost, per bridge that is measured is very high during the developing stage. In the implementation phase, the number of measured bridge decks will have to increase substantially. Yet, the cost per bridge deck will still be rather high, since measurement and interpretation productivity increases very slowly in this phase.

It is very tempting to use maintenance funds for new expensive measurement methods if there are potentially expensive maintenance problems that can be solved. There is a conflict here, since the necessary verification measurements on "reference populations" will still have to be financed by R&D budgets. Since R&D budgets constantly are under pressure, there is a risk that reference measurements are not made at all.

16 REFERENCES

Imregun, M. & Ågårdh, L 1993. *Impact excitation and updating of the FE model*. SP TN 1993:66, Borås: SP.

Swedish National Road Administration, 1996 *Bridge Inspection Manual* Publ 1996:036 (E)

Ulriksen, P. 1992 *Multistatic radar system - MRS* In P. Hänninen & S Autio (ed.) *Proc. 4th Int. Conf. on Ground Penetrating Radar June 8-13, 1992, Rovaniemi Finland*. Geological Survey of Finland, Special paper 16, Espoo 1992

Ulriksen, P., Wiberg, U. 1997. *Automatisk avsökning av betongkonstruktioner med utrustning baserad på mekaniska vågor* Elforsk Rapport 97:23 (in swedish, english summary) Lund University of technology

Wiberg, U. 1993. *Material Characterization and Defect Detection in Concrete by Quantitative Ultrasonics*. TRITA BKN Bulletin 7 Stockholm: Royal Institute of Technology.

Ågårdh, L. 1995. *Tillståndsbedömning av betongbroar genom mätning och beräkning av dynamisk respons*. (in Swedish) SP-AR 1995:31, Borås: SP

Operation and Maintenance of Large Infrastructure Projects, Vincentsen & Jensen (eds)
© 1998 Taylor & Francis, ISBN 90 5410 963 7

Reliability centred maintenance of equipment and systems

Geir Langli
The Norwegian Maritime Research Institute A.S. (MARINTEK), Trondheim, Norway

ABSTRACT: This paper discusses the application of Reliability Centred Maintenance (RCM) to equipment and systems from a design and operations point of view. Failure rates and dominating failure modes often vary with time and the consequences of a specific failure may also vary with time due to varying operating conditions. Furthermore, adequate reliability data is not always available. In addition, incomplete knowledge of the interaction between systems and selection of improper tools and methods for maintenance may contribute to deficiencies in the maintenance programmes. Maintenance strategies for equipment and systems are, for these and other reasons, therefore subject to revisions over time. It is a challenge to establish a decision basis to support revision of maintenance programmes. In addition to provide a brief introduction to the RCM analysis process, the paper presents experiences with and results of RCM analysis. An approach to continuous improvement of maintenance is also presented.

1 INTRODUCTION

The first preventive maintenance (PM) programme developed based on the principles of RCM was for the Boeing 747. This was back in 1968. PM programmes for DC-10, Concorde and Airbus A300 followed shortly after. The RCM concept was introduced to the US Air Force and US Navy in the early 1970s, and to the RAF in the early 1980s. The US nuclear industry began to investigate the potential of RCM in the mid 1980s. In the late 1980s, RCM was introduced to the Norwegian oil & gas offshore industry, and other industries have followed. In 1994, MARINTEK introduced RCM to the Norwegian shipping industry through the research programme *Information Technology in Ship Operation* (MITD).

When RCM was first applied by airlines, significant reductions in PM man-hours, whilst increased regularity and improved reliability were reported. The great savings in maintenance cost combined with improved performance made it easy to transfer the method to other industries. We now know that all of the promised and expected achievements did not materialise. We also know why. The challenges related to implementation were too often over-looked. Furthermore, whereas aviation possessed substantial experience in terms of failure and maintenance data, other industries did not. Lack of reliable reliability data, in terms of failure modes, failure rates and failure rate distributions, as well as incomplete knowledge of interactions of complex systems can severely bungle the performance and benefit of an RCM analysis.

In order to understand the break-through for application and development of reliability technology in aviation it is worth mentioning two early findings:

1. Scheduled overhaul has little effect on the overall reliability of a complex item unless it has a dominant failure mode.

2. There are many items for which there is no form of effective maintenance.

Figure 1 shows the various types of failure rate distributions that United Airlines identified. Figure 1 also includes the corresponding data from US Navy. These failure rate patterns clearly indicate why time-based maintenance fails so easily. The results from United Airlines showed that scheduled overhaul/replacement would have no effect on 89% of the failures. The corresponding figure for US Navy was 77%. This is because the failure rate is (more or less) constant, i.e. independent of time. This explains the problems aviation experienced. It is not possible to prescribe a PM programme for an item that very well may fail tomorrow as in N years. The only way to handle components and systems with such failure rate characteristics is by prescription of a condition-based maintenance strategy.

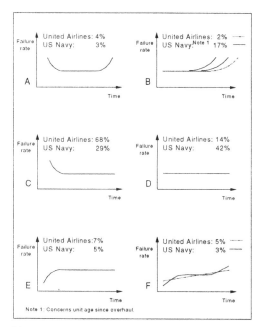

Figure 1. Failure rate distributions.

Since 1987, MARINTEK has been involved in setting up, performing and conducting RCM analyses, including development of RCM procedures, training and development of computer tools for RCM analysis. The following discussion is based on MARINTEK's experiences with R&D work in maintenance including RCM. Key elements in this context include integration of RCM analysis functionality in Maintenance Management Systems (MMS) and continuous improvement.

2 MAINTENANCE OF EQUIPMENT AND SYSTEMS

According to ISO/DIS 14224 and IEC 50(191), maintenance is "the combination of all technical and administrative actions, including supervision actions, intended to retain an item in, or restore it to, a state in which it can perform a required function".

In accordance with this definition, maintenance include the following elements, often referred to as the maintenance function:

1. Strategies for maintenance management, including objectives and control parameters.

2. Systems for planning, management and control.

3. Required procedures and documentation for planning, performance, reporting and analysis of maintenance activities.

4. Tools, materials and equipment.

5. An organisation that fulfils the stated requirements to competence and flexibility

RCM is about strategies, and strategies affect operations and personnel. It is therefore necessary to be aware of the factors that influence maintenance.

2.1 *Organisation of maintenance*

Maintenance strategies must be based on a maintenance philosophy that is compatible with the business objectives and the operations philosophy. The maintenance philosophy should state whether organisation of maintenance should be based either on centralisation, decentralisation or a combination of the two, and if certain tasks should be placed on contract (out-sourcing). The maintenance organisation should clearly define the responsibilities of various maintenance personnel and operators (e.g. for 1. line maintenance), and the competence required.

The maintenance programme should include guidelines for the use of maintenance management systems (MMS), condition monitoring (CM) systems, training programmes and mechanisms for experience acquisition. The maintenance organisation should define the relevant disciplines, lines of communication, names and responsibilities.

The RCM procedure is based on this type of information, and the results of an RCM analysis will often concern many of these aspects.

2.2 *Maintenance in relation to risk and RAM*

In accordance with the definition, maintenance is needed to ensure availability, and is thus also directly influencing - and influenced by - the reliability of an item. The worse the reliability, the more maintenance is required. The inherent availability A_i and the operational availability A_o of a system can be defined as in Equation 1 and Equation 2 respectively:

$$A_i = \frac{MTTF}{MTTF + MTTR} \tag{1}$$

$$A_o = \frac{MTBM}{MTBM + MDT} \tag{2}$$

Mean Time To Failure (alternatively, Mean Time Between Failures) and to some extent also Mean Time To Repair are hardware related, whereas Mean Time Between Maintenance and Mean Downtime depend on the maintenance function and logistic support etc. Reliability, Availability and Maintainability (RAM) are thus tightly connected to RCM.

A failure may result in downtime and it may also result in hazardous situations and accidents. It is thus important that safety functions and systems/equipment with inherent risks are properly maintained. However, the relation between maintenance and safety has a duality. Maintenance is a prerequisite for availability of e.g. production systems and safety systems. On the other hand, many disasters in various industries were directly or indirectly caused by deficiencies in the maintenance function, including organisational, managerial and human failures related to maintenance. An example is Piper Alpha. The cause-mechanism included failures in maintenance administration, reporting of maintenance, and in the performance of maintenance actions.

2.3 Maintenance and quality

Maintenance is also a prerequisite for quality. A production system or a transportation system cannot provide its product in accordance with specifications unless the production systems and the organisation function as intended. It is thus necessary to monitor the performance and technical condition of systems and equipment to provide the required level of maintenance. Quality assurance of a production process and its products should inform adjustment of maintenance strategies.

2.4 Maintenance strategies

Maintenance philosophies may be categorised as either pro-active or repair-based, which strictly speaking is an expression for lack of philosophy. Even though the repair-based approach prevailed up to World War II, there are still corporations applying this philosophy. A study of US refineries for the period 1986 – 1992 (Ricketts, 1994) showed that the pro-active (reliability-based) refineries could demonstrate a far better performance than the ones that applied a repair-based approach. Whereas the 11 refineries with the largest cost increase had an average increase in the maintenance index from 12.5 to 30 ($/EDC) and a decrease in mechanical availability from 95.6% to 93.7%, the corresponding numbers for the nine refineries with the greatest improvement were 22/20 ($/EDC) and 92.2%/95.8%. (EDC is parameter reflecting refinery capacity and complexity.)

In order to assign the correct maintenance strategy for a given piece of equipment or system, the factors influencing the maintenance must be identified. It is necessary also to be aware of the alternative maintenance strategies and methods that exist. Maintenance strategies can be grouped as shown in Figure 2. Unplanned corrective maintenance is not referred to as a strategy since this

implies a fault in the PM (time-based maintenance (TBM) / condition-based maintenance (CBM)) - programme. Planned corrective maintenance (PCM) is also referred to as breakdown maintenance (BDM).

There are several methods available for establishment and review of the maintenance function, including development of maintenance strategies, such as Total Productive Maintenance - TPM, Risk Based Inspection - RBI, Risk Based Maintenance – RBM, and RCM.

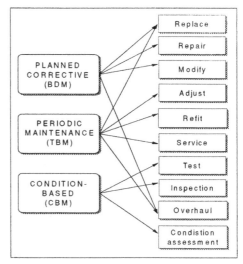

Figure 2. Maintenance strategies.

2.3 Characterisation of today's maintenance

Maintenance is steadily attracting more and more attention. Reasons for this include cost cutting and competitiveness. Advances in information technology and surfacing of new management concepts like Total Productive Maintenance (TPM), Total Quality Management (TQM) and Business Process Re-engineering (BPR) are also important, both as methods and as catalysts. From being perceived as a cost, maintenance is now regarded as an investment. Managers are conscious of the fact that maintenance accounts for a substantial amount of overall operations costs and that there is room for improvement (i.e. cost reductions). At the same time, managers are fully aware of the need for optimum maintenance in order to stay in business. A result of this change is that MMS manufacturers and suppliers report record sales.

Characterisation of maintenance and maintenance performance of Norwegian maritime and offshore industries is not easy, because the industry is diverse. There are a number of different MMS around. A ship owner with 10 ships may have 10 different MMS', old ones as well as new ones. This is partly because the ship owner may leave the yard to purchase the MMS, but mainly because ownership will, when buying second-hand, also inherit the vessel's MMS and he seldom perceives it as cost-effective to install a new system. Maintenance has been decentralised, resulting in inconsistent performance and poor reporting of maintenance. In addition, crews have different educational, cultural and professional backgrounds. All together, this makes control of the maintenance function a very demanding task. Man-hours for maintenance has not been logged against tagged items, only in those cases where maintenance have been performed by contractors. Likewise, control of spares has also suffered. In such an environment, control and follow up of maintenance is hard. Despite this, availability and regularity has been quite good.

Just recently, a few Norwegian ship owners have initiated a process with the objective of centralising major parts of the maintenance.

The use of MMS is not, and has never been, satisfactory. This is due to several factors. However, one of the most important factors is an almost complete lack of feedback from the land-based organisations. As part of the research project *Teknisk Tilstand* (Technical Condition), a study has been performed on the quality of MMS and their use with respect to collecting data that can be used for studying the relationship between the influence factors (given by operating conditions) and degradation mechanisms and degradation processes. *Teknisk Tilstand* is a joint industry project also sponsored by the Research Council of Norway (www.marintek.sintef.no/mt23doc/tekn-til).

The study concluded that MMS' in general were not used in accordance with intentions, and the quality of use varied between shifts and over time. Data contained within the MMS' were not suitable for analysis of cause-mechanisms of failures and effects of maintenance. The MMS' do not support such data collection, and because of lack of feedback users adopted their own particular ways of operating the MMS. These findings were in general applicable to all companies in the project, and, the problems were directly comparable to those of shipping and rig operations.

Improper MMS' and improper use of MMS are thus industry independent, even though deviations are less frequent and less safety critical in the high-risk industries. The root problem is that maintenance management is performed as maintenance administration, and this is because organisations at large have no or only incomplete mechanisms for analysis of operations and maintenance data and continuous improvement of its business. Personnel responsible for maintenance have suffered from the lack of proper methods and tools for justifying the

need for and prioritising of maintenance. Some engineers, operators and managers of organisations that have applied RCM, perceived RCM as a pill that once swallowed the ache would disappear. The challenges in connection with implementation of the results of RCM have been overlooked, or at best, underestimated.

In 1991/92 MARINTEK was involved in RCM analysis of one of the major Norwegian oil & gas installations. The analysis concluded that PM could be reduced with an average of 15%. In some cases, PM had to be increased, whereas the extreme on the other side was a reduction in PM of 82%, see Table 1. The client was happy.

A year after completion of the analysis, platform management asked where the savings were. The maintenance costs did not decrease, they tended to increase. There were two reasons for this. Firstly, there was no system in place for actually logging and monitoring maintenance costs versus system. Secondly, an ever changing scene of changes in operations and modifications to systems and equipment constituted changes in the conditions underlying the RCM analysis, but the effects of these changes were not - and could not - be identified. The problems were related to the use and functionality of the MMS.

Table 1. Extract from results of RCM analysis of a major Norwegian offshore oil & gas platform.

SYSTEM	PM MAN-HOURS			
	before	after	change	%
UB/A Cooling Medium	760	798	+38	+ 5
UB/B Cooling Medium	100	132	-32	+ 32
UB/C Cooling Medium	214	94	-120	- 56
PG/B Gas Coolers & Scrub.	631	278	-353	- 56
PG/C Gas Coolers & Scrub.	838	770	-67	- 8
PE/A Crude Metering	411	526	+115	+ 28
PE/B Crude Metering	382	256	-126	- 33
AF/A Space Heat Vent	2450	1764	-686	- 28
AF/B Space Heat Vent	3291	2534	-757	- 23
AF/C Space Heat Vent	1080	972	-108	- 10
UC/A Steam & Hot Water	305	176	-131	- 43
UC/B Steam & Hot Water	89	81	-8	- 9
UC/C Steam & Hot Water	152	28	-124	- 82
WA/A Wellhead & Manif.	6700	6633	-67	- 1
WA/B Wellhead & Manif.	2200	2156	-44	- 2
WA/C Wellhead & Manif.	3117	2930	-187	- 6
BF/A Material Handling	2230	870	-1360	- 61
4 other systems	865	868	+3	- 0
SUM	25814	21863	- 3951	- 15,3

The ultimate goal of the RCM concept is to identify the required type and level of maintenance, no more and no less, to support life-long operations. Implicitly, this is about condition-based maintenance (CBM). The use of condition monitoring (CM) systems is thus worth commenting upon. The size of the «condition monitoring congregation» has varied over time, reflecting the beliefs and disbelief in CM systems. Suppliers of CM systems may have promised more than they should. On the other hand, few organisations have managed to implement CM because they did not know what it would take. We have seen gas turbines equipped with expensive CM systems that have been maintained based on a conservative time-based maintenance strategy. In such cases it is difficult to justify the investments in CM. We have also seen examples where plant owners have selected expensive CM systems where in fact quite simple systems and arrangements would have been sufficient. The choice of CM does not reflect the actual needs – these are not always are well understood and documented, and few organisations are actually aware that introducing CM has organisational implications.

RCM is now accepted as a method for the establishment of initial PM programmes and optimisation of existing PM programmes. We have observed a significant change in the attitude towards the use and implementation of RCM analysis. This surfaces in two ways. Expectations are more realistic and diverse, and they are linked to the need for improvement of the maintenance function, review of spare parts, and securing of competence and knowledge. The latter is a consequence of de-manning due to introduction of new technology and revised work processes. In the Norwegian oil industry, RCM may become a statutory requirement. Secondly, the industry is concerned with the interaction between RCM and MMS, and how to integrate these.

The Norwegian oil and gas industry is now in a process that may result in that all oil companies will use SAP R3 as the main operations system. SAP is an integrated system, and includes a MMS function. The industry disagrees, however, to what extent the functionality provided by this system and similar systems is adequate with respect to their needs regarding maintenance management. Such systems are introduced in companies mainly for other reasons than maintenance, hence MMS is just one piece in the jigsaw puzzle that makes up an integrated operations system. It is thus important that personnel responsible for maintenance are aware of the requirements they have to meet in the pre-implementation phases for such systems.

The international standardisation efforts will affect maintenance in many respects. This concerns the use of product data models for sharing and exchange of information, in design and operation. This will also affect maintenance in all aspects and all phases of a project, from design through operation to abandonment. It is important that people engaged with maintenance join forces to make the best use of these developments.

3 RCM IN A LIFE CYCLE PERSPECTIVE

In aviation, RCM is considered as a life cycle process, and systems and routines are in place to support this philosophy. Other industries have also recognised that RCM analyses need to be reviewed in order to incorporate experience. However, offshore oil and gas, shipping and other industries lack systems and routines to realise such a philosophy.

Optimisation of maintenance is an activity that attracts attention some years after commencement of operations. When in operations, such optimisation can only result in sub-optimisation. In order to engineer an installation that can be run with minimum maintenance, maintenance must be focused in the design phase. The inclusion of structured maintenance engineering activities in design can effectively contribute to identification and removal of technical solutions that have unacceptable maintainability characteristics. Further, maintenance engineering can provide useful input to manning studies and to the identification of the need for, and the specification of, CM systems and MMS. Finally, such activities will also provide essential input to the evaluation of spare part requirements and the recommendations of manufacturers in this respect.

A carefully defined RCM analysis can be very efficient tool in this respect. This includes to select the systems to be included in the RCM analysis.

3.1 RCM in the design stage

The maritime and offshore industries have been reluctant to apply RCM in the design phase of a project. This is partly because the potential of RCM is not fully recognised by most engineers in the design teams, partly because of tight schedules, partly due to the objective of minimising expenditures. RCM analysis is a time consuming and resource demanding exercise.

It is not feasible or necessary to perform a full RCM of all systems in the concept phase. The systems subject to analysis should be those that are critical from a production or safety point of view, i.e. those systems for which PM will be assigned. In addition, one should focus on those systems which are maintenance intensive.

Use of simplified RCM can be an effective tool to:

1. eliminate design solutions that would require PM,

2. ensure early identification of potential/ impending failures (design for on-condition maintenance and damage tolerance),

3. eliminate support structures/work platforms for both PM and CM, or identify the need for such structures,

4. identify and eliminate hidden failures of major equipment units/packages, and

5. support maintainability studies

In detail engineering and procurement, a full RCM should be performed of critical equipment and maintenance intensive equipment, e.g. valves, piping segments, rotating machinery and structures. RCM in this phase of the life cycle of an installation is recommended in order to:

1. Establish initial maintenance programme
2. Evaluate alternative maintenance philosophies
3. Provide input to/interface maintainability

study (e.g. test points, sampling points, accessibility)

The Norwegian Petroleum Directorate (NPD, 1997) has in a report, emphasised the above and states there is a need to focus on maintenance in the design phases of offshore installations.

3.2 RCM in the operational phase

In the operational phase the areas of application of RCM analysis include review/update of PM programmes (PM/PCM ratios, adequacy of maintenance strategy and maintenance and inspection intervals), review/update of spare part stock levels, and decision support in connection with planned annual maintenance shutdown.

Review of the RCM analysis in operations should be performed on a regular basis and when performance implies a need for such a review. Reviews should also be performed when changes are introduced in the boundary conditions that influence maintenance, i.e. hardware, operations philosophy and operating conditions.

The main purpose of reviewing and updating the RCM analysis is to incorporate experience related to failure modes and failure rates of equipment and systems. This is necessary in order to justify changes to maintenance strategies (task and interval) for equipment and systems.

The maintenance strategies are based on the objective of ensuring life-long operation of equipment and systems. When the operational lifetime of equipment, systems or an installation comes to the end, there is a need for review of the maintenance programme. Significant cost reductions can be achieved by cancellation of maintenance activities and extending intervals, without impairing safety and availability.

3.3 Examples on RCM analysis results

A common finding of RCM analyses is that the existing maintenance strategies do not reflect the actual operating conditions of equipment and systems. An example is where e.g. a control valve in a fresh water system compared to a situation where it is part of a sea water system. The dominating

failure modes might differ and the failure rate would be higher for the one in the sea water system. Furthermore, the maintenance demand would be more intensive if the same control valve was part of system containing hydrocarbons. This because the consequences of a failure has a greater accident potential. This is all quite obvious, but when defining the initial PM programme for a plant it is not possible to deal with all these aspects without a structured analysis method like the RCM method.

An RCM analysis performed *for Mærsk Olie og Gas* in 1996 in the construction and commissioning phases for the offshore platforms *Svend, Tyra East* and *Tyra West* resulted in substantial annual savings in spare parts per platform. The savings obtained for one year for either module were almost twice the overall cost of the RCM analysis.

A study of the gas compression train for *Norsk Hydro's Brage platform* (1992) revealed a need for design modifications to the gas turbine. These modifications concerned instrumentation, control valves and the fuel system.

In 1994, *MARINTEK* performed an RCM analysis of the *Draugen* platform operated by *A/S Norske Shell*. A summary of the analysis of 40% of the critical sub-systems is presented in Table 2. The pay-back time for the analysis was three months.

Table 2. Extract from the results of the RCM analysis of *Draugen*. Note: The PM man-hours relate to direct maintenance work only.

| SYSTEM | PM MAN-HRS/YR. | | CHANGE | SAVINGS | RCM COST |
	before	after	(%)	(US$/yr.)	(US$/yr.)
Oil	1.927	918	-52	78.400	20.000
Gas/cond	2.138	1.765	-17	29.000	22.000
Electrical	2.859	2.386	-17	36.900	86.000
Fire&gas	11.221	3.574	-68	594.000	57.000
Fire prot.	1.173	986	-16	14.500	4.000
HVAC	1.176	344	-71	64.600	10.000
Cranes	523	263	-58	28.000	5.000
Seawater	580	238	-59	26.500	4.000
Water inj	619	173	-72	34.600	5.000
SUM	22.316	10.647	-52	906.500	213.000

We have also performed RCM analysis for shipping companies that did not result in any change to the PM programme. However, in this case the analyses have served to document the maintenance programme, to acquire the experience and to form a basis for continuous improvement.

4 THE MAIN STEPS OF AN RCM ANALYSIS

4.1 The RCM process

An RCM analysis may result in considerable changes to the maintenance function. The analysis should therefore be performed by those who are involved in operation and maintenance of the systems subject to RCM analysis. Ownership of both the analysis process, the input data and the results are of utmost importance. Over the years, MARINTEK's role has changed from performer to facilitator. This role includes preparing, in co-operation with the client, the RCM procedure, providing the basic training in application of the method, and functioning as chairman/catalyst during the analysis period.

RCM analysis is teamwork, and the team should possess the following qualifications:

1. Experience from operation and maintenance of the plant, or similar plants or systems

2. Knowledge of the design premises for systems and equipment and the intended functions

3. Knowledge of actual and possible failure modes, failure frequencies and consequences of failures

4. Knowledge of failure characteristics

5. Knowledge of manufacturers, documentation and history (for existing plants)

6. Knowledge of resource requirements (personnel, spares and materials) for maintenance

7. Knowledge of relevant HSE regulations and requirements

8. Knowledge of the RCM procedure

In order to carry out the analysis, the following background information should be available:

1. System descriptions and equipment lists

2. Manufacturers drawings and documentation

3. Plant general arrangement and equipment layout drawings

4. Piping and instrumentation diagrams

5. Safety studies and material handling studies

6. Failure / event history (for existing plants)

7. Spare parts history (for existing plants)

8. Maintenance statistics (for existing plants)

Important success factors for an RCM analysis are management commitment, active and continuous participation of end users, and sufficient time.

4.2 The RCM procedure

The key to a successful RCM analysis begins with the establishment of the RCM analysis procedure. The procedure should state the maintenance policy, e.g. scope of 1. line maintenance, where the maintenance is to be performed, who shall perform the work, and the parameters used for monitoring and control of operations. If a certain policy applies to certain equipment, e.g. transformers are be maintained by a contractor, this should also be included in the RCM procedure.

The RCM procedure defines the input and output requirements, the criteria to be used for criticality analysis, and provides the decision tree logic for assignment of maintenance strategy. The procedure may also include a decision tree for establishment of intervals.

4.3 Function analysis

The purpose of the function analysis is to establish the functional relationships between the various function levels, i.e. a function hierarchy. This task is mentally quite challenging, because most engineers and operators are used to thinking of systems and equipment in a physical context.

The breakdown of functions into sub-functions terminates at the level of maintainable items. If a piece of equipment is to be serviced by a third party, it may be sufficient to terminate at equipment or package level. If internal maintenance personnel are responsible for the maintenance of an item, it may be necessary to continue the breakdown to the level of components and parts.

When the function hierarchy is established, the analysis of function failures commences. A function failure can be a complete or partial loss of function. In order to limit the work, a function failure should be significant with respect to frequency of occurrence or its potential consequences.

An important part of the function analysis is to decide whether the function failure is hidden. A consequence of a hidden failure is an increased risk for serious events.

4.4 Failure Mode and Effect Analysis

The first step is to perform an equipment analysis. The purpose of this analysis is to link equipment failure modes to function failures. Since equipment normally consists of several parts, there may be several failure modes. One should thus focus on the significant function failures. The causes of the identified failure modes should then be identified, together with the method for detection of the actual failure modes. Detection methods can include condition monitoring, function testing, visual inspection, to mention some of the most common ones.

Furthermore, the FMEA should include MTTF values. In many cases it may not be possible to find the required values from available reliability databases and data handbooks. The desired figures may simply not be present, or the context and technological aspects may not be relevant. This is the point in the analysis where the experience of operations and maintenance personnel is invaluable. They may not be capable of quantifying the MTTFs etc., but when «forced» to think in lines of «once a month, once a year, once every 10 year» it usually possible to arrive at reasonable values. The criticality classification scheme is normally coarse with respect to MTTF. The need for good MTTF figures is therefore only required for PV/TBM internal estimation purposes.

4.5 Criticality analysis

The objective of the criticality analysis is to decide whether preventive maintenance is recommended. The criticality analysis commences when the equipment analysis (FMEA) for a particular equipment/system is complete. Criticality, in the context of an RCM analysis, is almost synonymous to risk. Criticality is to be perceived as a measure of the risk associated with an equipment failure. The criticality analysis is the combined evaluation of the frequency and the consequences of a function failure. The consequences of a failure are evaluated with respect to a set of risk categories, for instance:
1. personnel safety
2. environment
3. availability
4. cost

The criticality analysis may be performed qualitatively, but some kind of quantification is recommended. An example of a consequence matrix and a criticality matrix are presented in Figure 3 and 4, respectively.

With respect to availability, redundancy means that the criticality may be less severe. With respect to safety and environment, redundancy means a potential increase in criticality if the systems/ equipment in question contains hazardous substances. For safety systems and non-critical systems redundancy implies potential decrease in criticality.

If the overall criticality is higher than '2' (Figure 4) for one or more of the stated parameters, the equipment failure is considered critical and PM shall be considered. Whether PM actually can be prescribed is dependent of the existence of an available method that is effective and cost-effective. A criticality of '2' or lower means the equipment failure is not critical, and the equipment may be assigned planned corrective maintenance (PCM).

Figure 3 and Figure 4 represent approaches to RCM for industrial domains where reliable /applicable reliability data cannot be identified, or where the end users are reluctant to accept available data as «the truth». As long as the end users can relate their own knowledge to both consequences (type and severity) and frequencies of occurrence, it is easier to implement the outcome of the analysis. However, to base the analysis of some kind of semi-quantitative approach is not the same as saying numbers are of no value. The RCM analysis facilitator should be well prepared and familiarised

Risk Category	Consequence		
	Negligible (N)	Significant (S)	Catastrophic (C)
S	None	Personnel injury	Fatality
E	None	Moderate releases with temporary and restricted environmental consequences, or breaches to statutory HSE regulations	Major releases or severe breaches of statutory HSE regulations.
A	Downtime < 1 hours	1 hour ≤ downtime <3 hours	3 hours ≤ downtime
C	Total cost of CM is less than total cost for PM	Total cost of CM is larger than total cost for PM	Total cost of CM is 15 times l arger than total cost for PM

Figure 3. Consequence matrix.

Prob. of top event	Consequence				
	Negligible (N)		Significant (S)		Catastrophic (C)
Probable event	3 *)	S ↓ M ↓ K ↓	4		5
Event has occurred with the company	2		3		4
Low probability, but the event has occurred	1		2**)	S ↑ ↑ ↑	3
Improbable event	0		1		2

*) Value for safety, environment and cost reduces to 2.
**) Value for safety increases to 3

Figure 4. Criticality matrix.

with relevant data in order to guide the discussion into the right tracks.

4.6 Identification of maintenance strategy

FMEA and FMECA have been around for many years. As such, one might argue that the RCM analysis begins when the FMEA is completed. It is the decision tree that is used for identification of maintenance strategies that is perceived as the core of an RCM analysis.

The decision tree must reflect the governing policies and characteristics of the functions/systems in question. E.g., a decision tree for military systems will differ from that of a drilling rig, and a complex plant will require a different decision tree compared to a plant made up of relatively simple systems and equipment. There exist decision trees that can serve as some kind of branch standard, reflecting maintenance philosophies. Aviation has its trees (Nowlan & Heap, 1978, MSG-3, 1980), the nuclear power industry also has its own (Hoch, 1990), and more generic decision trees are available (Moubray, 1991). MARINTEK has applied a decision tree for shipping and is about to apply a slightly different tree for rig owners, see Figure 5. Each industrial branch has its characteristic features that enable some degree of standardisation of the RCM analysis procedure and hence decision trees. However, some aspects are tightly coupled with the companies' business idea, technological status and operations philosophies. Identification of maintenance strategy, e.g. CMB, TBM or PCM, is good, but in addition one must determine how often a maintenance task is to be performed.

Reliability data, which constitutes the basis for the estimation of intervals, has its limitations. Reliability data is strictly speaking only valid for the equipment they were collected from. Different designs, different materials, different manufacturing processes, different operations and maintenance philosophy etc. introduce large standard deviations to equipment failure rates. There might be over two order of magnitudes between the lower and upper confidence limits.

Guessing at maintenance and inspection intervals can therefore be a difficult task, even though there are rules of thumb and theoretical models in place that can be effective aids in this respect. In some cases authorities have defined inspection and test intervals, in particular for safety-related equipment like fire pumps, detectors and ESD valves.

Establishment of intervals is easily supported by a decision tree logic. Questions that should be asked include whether the function failure is hidden and whether the function failure is critical with respect to safety or the environment. Intervals are also dependent on degree of redundancy and what is considered an acceptable unavailability.

The optimal solution will be based on a mix of different needs and requirements:

1. Test interval to reveal hidden function failures
2. Inspection interval for condition assessments
3. Fixed age/fixed time intervals to comply with statutory safety and environment requirements and to reduce costs.

An RCM analysis can further include analysis of manning requirements (disciplines and man-hours for active repair) for the identified PM and TBM actions. Maintenance means downtime, and it will further be of interest to identify the total downtime, i.e. both active repair time, logistics delay time and administration time.

Some maintenance will necessitate shutdown of the production system. With respect to job packing and planning of maintenance, it is wise to include this assessment as part of the RCM analysis.

Finally, the RCM analysis provides a framework for evaluation of spare parts requirements. Whether to procure complete units or unit parts, number of spares, actual and allowable lead-time requirements, and where to locate spare parts may also be included in the RCM analysis.

4.7 Job Packing

The results of an RCM analysis are of little value if not implemented into the MMS. Implementing the

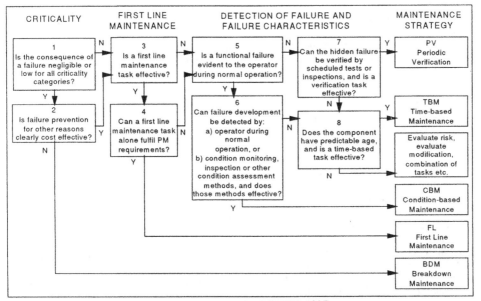

Figure 5. Decision tree logic for mobile drilling units (MARINTEK, 1997).

results without some kind of post-processing is not possible. The jobs need to be sorted by interval, system and whether shutdown of the production system is required. The jobs can easily be input to the MMS provided the analysis and job packing is performed using a computer-based tool.

5 CONTINUOUS IMPROVEMENT OF THE MAINTENANCE FUNCTION

5.1 Demonstrator of continuous improvement

The project *Optimal Maintenance and Spare Part Requirements* in MITD revealed the need for data-assisted support for continuous improvement of the maintenance system. This tool should further be capable of analysing large amounts of historical data and guide the operator's attention to areas that deviated from the plan. Through discussions with ship owners a total of nine control parameters were defined. The three control parameters listed in Figure 6 are the ones that were selected for the prototype development. The demonstrator was to be implemented in a MMS.

The concept on which CIM is developed is shown in Figure 7.

The demonstrator called *CIM* - Continuous Improvement Module. The MMS *RAST* by *MARINOR* was used as platform for the demonstrator, and the model was tested on real data. For demonstration purposes, the following control parameters were focused upon:

1. condition before maintenance
2. unplanned corrective maintenance
3. spare part consumption

Normally, the condition after maintenance is logged. With respect to maintenance intervals and maintenance tasks, it is interesting to know what the condition prior to the maintenance is. If the condition before maintenance repeatedly is good (degree of acceptable condition), this implies that the interval is conservative.

If, on the other hand, the condition before maintenance is unacceptable (close to breakdown), this implies that the interval should be increased, or, that the actual maintenance task is inadequate. Likewise, if the number of unplanned corrective maintenance is higher than what is deemed acceptable, then there is reason to suspect that the maintenance strategy is inadequate.

To each of the control parameters' limit values were defined. These were used as checkpoints to filter the historical data. The control parameters presume that condition is reported using the class societies' CAP system (score from 1 (as good as

No.	Control parameter	Deviation	Deviation limit
1	Technical condition assessment before maintenance job is started	Technical condition before the maintenance job is started is too good	Technical condition is better than CAP 2 in 80% of the cases
2	Unforeseen failure occurs	Failure has critical consequences	Any case occurs
3	Spare part usage	Spare part is not used	Time in stock > 2 times the planned maintenance interval (for periodic maintenance jobs). Time in stock > 2 times the ETTF (for CBM and PCM jobs

Figure 6. Examples on control parameters used in CIM.

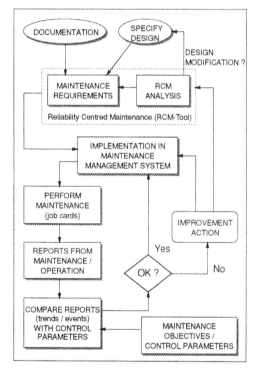

Figure 7. Concept for continuous improvement of maintenance.

new) – 4 (significant deficiencies - maintenance required)).

In the CIM, three parallel tests are performed on the control parameter condition before maintenance:

Condition before maintenance = CAP 1 in X% of N cases
Condition before maintenance = CAP 2 in Y% of N cases
Condition before maintenance = CAP 4 in Z% of N cases

CAP 3 prior to maintenance is deemed a proper condition before maintenance, hence cases that have been reported as CAP 3 are not evaluated. The limit values are user-defined, and several sets of limit values (sets) can be defined. There must be minimum number of historical records (cases) for each maintenance job to execute the test.

The maintenance tasks in the MMS subject to analysis are connected to the relevant sets. One particular maintenance task can be connected to several sets. Prior to execution of the test, the user must select a number of tasks from the account system of the MMS. The analysis is then performed on all tasks registered with a particular account number. That is, each job in the maintenance system connected to one or more of the control parameters are analysed with respect to all the connected control parameters. The history of each analysis is recorded.

Each analysis provides a list indicating the job number, control parameters and the result of the analysis for each hit. A dialog box is provided that also includes a facility for changing the interval. If the interval is changed, the job card in the MMS is automatically updated. When an interval is changed, this is logged in the MMS together with old value and date. The analysis based on condition before maintenance has not been completed by the time of preparing this paper, due to the fact that condition before maintenance is reported in the MMS.

The test performed on spare part consumption revealed that control was poor. Data from a Norwegian chemical tanker was used. For a particular item the minimum number of spare parts on stock should be 1. The order level is 6 and is established based on assumptions on delivery time and rate of consumption. The analysis showed that the number of spare parts at any time was below 4, and that maximum number of spares on stock was 12. The consumption of the particular spare was insignificant between planned maintenance. The ship owner concluded that an average reduction of 20% was achievable.

Furthermore, of the spare parts valued at NOK 7.800.000 onboard a 30.000 dwt tanker, 67% had a unit cost of NOK 5.000 or less. Spare parts with a unit cost of NOK 250 or less counted for 27% of the total value of spare parts onboard. This was a surprise since the general idea of spares and cost was that it was the expensive units that accounted for most of the cost. One of the conclusions of the analysis was that there is reason to optimise on low-cost spare parts.

The demonstrator has so far proved a great potential for spare parts stock optimisation and for identification of unplanned corrective maintenance.

5.2 *RCM: A vehicle for continuous improvement*

CIM should be a functionality that is accessible from the MMS. The same recommendation applies to RCM. So then, what is the connection between CIM and RCM?

Suppose that the RCM tool supports the maintenance analysis through to job card development and job packing. The output of the RCM analysis can then be transferred to the MMS with no further processing. Whereas the MMS only contains the job cards and the maintenance plan (tasks and intervals), the RCM analysis contains the rationale behind the plan. When the CIM identifies a deviation, e.g. the condition before maintenance is better than the acceptance limit, the operator can access the RCM tool from the MMS to find out about why the particular interval was defined. Likewise, if the number of unplanned corrective actions exceeds the pre-defined acceptance level, the operator may access the RCM tool and check why the maintenance strategy was assigned in the first place. By revisiting the decision logic (Figure 5), the operator will easily see the rationale behind the strategy. The experience gained from operations and maintenance may gradually be used to revise the maintenance strategy.

The connection between CIM and RCM is thus beneficial from two points of view. Firstly, by conferring with the RCM analysis, one significantly reduces the risk of over-looking aspects that must be considered when establishing a maintenance strategy. Secondly, by updating the RCM analysis at the same time as the interval is changed the rationale behind the revision is secured for future use. A thoroughly performed and documented RCM analysis is an outstanding aid for training of future generations of maintenance personnel.

6 FUTURE ADVANCES IN RCM

There is a great potential for improvement of the RCM with respect to how evaluation of criticality is carried out, at least in the maritime and offshore oil and gas industries. The way risk (criticality) generally is dealt with is sufficient to capture experience. However, occasionally quantitative data and proper models of physical phenomena are available and ought to be used. The RCM framework can handle this kind of models, but to our knowledge, there is no computer tool available for RCM analysis that also provides an effective link between the analysis tool and such models.

Another challenge is to develop reliable models for establishment and modification of intervals for inspection, testing and repair activities. This is due to several factors, including quite rigid statutory requirements, lack of qualified reliability data and incomplete knowledge of the mechanisms that dictate the reliability performance of equipment and systems.

Whereas ageing mechanisms in aviation in general are quite well understood, ageing mechanisms still represent a challenge to most other industries. The objectives of the project *Teknisk Tilstand* include to identify the dominating degradation mechanisms for equipment, and to relate these mechanisms to influencing factors. These factors relate to operating conditions and design. In order to realise the full potential of RCM, the knowledge of what factors determine the lifetime of equipment or a system and how to monitor the long-term development in technical condition of equipment and systems is essential.

Because RCM is such a time-consuming and resource-demanding exercise there should be room for an application of a data warehouse approach to RCM analysis. Within a specific industrial domain, the systems, and hence functions, are identical or quite similar. In *Information Technology in Rig Operation* (RITD) (www.marintek.sintef.no/ritd), the rig owners have an interest in such an approach. The rig systems are distributed between the rig owners who then perform RCM analyses of these systems. The results are validated in plenary, and the other rig owners only adjust the analyses to reflect their specific requirements. The cost of performing RCM analysis is substantially reduced, the entire maintenance experience of a branch is contained within database, and consistency in maintenance performance and reporting is ensured.

Some MMS suppliers now offer some kind of RCM analysis functionality in their MMS', and there is a growing demand in the industry at large to take on RCM (or similar methods) as an integral part of MMS. The challenge in this context is to develop An RCM analysis tool that is sufficiently flexible to match the needs of various companies within different industrial domains. Of all the RCM analyses MARINTEK have performed over the years, there has been no copy-paste opportunity. As long as companies have their own policies, priorities and organisational set-up, maintenance programmes and hence RCM analyses will differ.

Provision of CIM facilities for MMS is a must in order to meet business requirements, including cost control and to support internal control. There is a need for further work into the areas of establishment of effective control parameters and design and operation of effective control mechanisms. Documentation about the basis for decision-making concerning maintenance is found in the RCM analysis, hence there is a mutual dependency between continuous improvement and RCM.

Finally, RCM also provides a vehicle for extracting and documenting the experience of operators and maintenance personnel. We have seen companies and corporations that have been forced to bring in retired personnel in order to finalise the planning of revision shutdowns and even participate in the actual work. There is an urgent need to capture the knowledge of experienced personnel for the next generation operators and maintenance personnel. Development of combined RCM analysis software that includes an interactive multimedia training facility may be a life buoy to many companies.

7 CONCLUSIONS

1. RCM has proved useful for establishment of preventive maintenance programmes in a number of industries. The method can be used in both the engineering and operational phases of project.

2. Application of RCM provides the right level of maintenance, and will normally result in less man-hours for PM and reductions in spare part stocks.

3. One main challenge of RCM relates to the availability of adequate data. However, RCM can effectively utilise qualitative information in terms of experience.

4. Another challenge is the implementation of RCM results in MMS. RCM analysis should be performed by the end users and with the aid of a computer-based tool. The MMS should further support reporting of maintenance work against maintenance control parameters such that follow-up and improvement of the maintenance programme is possible.

5. RCM is very well suited for extracting and cataloguing of experiences.

6. In order to realise the full potential of RCM in a life cycle perspective MMS' should be provided with a function for continuous improvement. This function should be linked to an RCM analysis facility in order to document the rationale behind revisions of the maintenance programme.

7. The application of a module for continuous improvement (CIM) together with a MMS has shown promising results with respect to spare part optimisation and to draw the operator's attention to areas that deviate from the plan.

LIST OF ABBREVIATIONS

A_i	Availability (inherent)
A_o	Availabiliy (operatioal)
BDM	Breakdown Maintenance
BPR	Business Process Re-engineering
CAP	Condition Assessment Procedure
CBM	Condition Based Maintenance

CIM	Continuous Improvement Module
CM	Condition Monitoring
ETTF	Estimated Time To Failure
FMEA	Failure Mode and Effect Analysis
FMECA	Failure Mode, Effect and Criticality Analysis
MDT	Mean Downtime
MMS	Maintenance Management System
MTBF	Mean Time Between Failure
MTBM	Mean Time Between Maintenance
MTTF	Mean Time To Failure
MTTR	Mean Time To Repair
PCM	Planned Corrective Maintenance
PM	Preventive Maintenance
PV	Periodic Verification
RAM	Reliability, Availability, Maintainability
RBI	Risk Based Inspection
RBM	Risk Based Maintenance
RCM	Reliability Centred Maintenance
TBM	Time Based Maintenance
TPM	Total Productive Maintenance
TQM	Total Quality Management

REFERENCES

Nowlan, F.S. and Heap, H. 1978. *Reliability-centred Maintenance*. National Technical Information Service, Springfield, Virginia, USA

Hoch, R.R. 1990. A Practical Application of Reliability Centered Maintenance. 90-JPGC/Pwr-51. *Jt. ASME/IEEE Power Generation Conference*, Boston, 1990

Ricketts, R. 1994. Organize to manage reliability. *Hydrocarbon Processing*, December, 51-54

Maintenance Steering Group – 3 Task Force: *Maintenance Program Development Document MSG-3*. Washington DC, Air Transport Association of America

MARINTEK, 1997. *RCM procedure for IT in operation of mobile drilling units*. Restricted.

Norwegian Petroleum Directorate (NPD), 1996. Maintenance Management – Experiences and Challenges (in Norwegian). OD-96-81. Stavanger, Norway.

Organization

Operation and Maintenance of Large Infrastructure Projects, Vincentsen & Jensen (eds)
© 1998 Taylor & Francis, ISBN 90 5410 963 7

Aspects of safety and operation of bridges during rehabilitation

M. H. Faber & D. L. Hommel
COWI Consulting Engineers and Planners AS, Lyngby, Denmark

R. Maglie
Dirección Nacional de Vialidad, Buenos Aires, Argentine

ABSTRACT: Bridge rehabilitation usually involves a broad variety of activities on the bridge site in addition to the verification and repair design work at desktop. As an integral part of the rehabilitation works inspections, tests, monitoring, repairs and strengthening are included. Besides the actual state of deterioration of the bridges and the corresponding residual load carrying capacity and service life, each of the rehabilitation activities may impose restrictions to the use of the bridge during the rehabilitation. Whereas inspections, tests and monitoring usually may be planned and conducted for periods of low traffic intensity in order to minimise the consequences to the operation of the bridge, repair and strengthening works are very difficult to conduct without some consequences to the operation of the bridge. To enhance a rational planning of repair and strengthening works, allowing for an optimisation of the operability of the bridge and at the same time ensuring a minimum safety during the rehabilitation works, a reliability based framework for rehabilitation is presented. The application of the proposed framework is illustrated on the basis of recent experiences from the rehabilitation of large cable stayed bridges in Argentina.

1. INTRODUCTION

To the owners and operators of bridges the rehabilitation and/or upgrading of the bridge capacity and service life is more than just identifying and executing an appropriate strengthening and/or repair design.

For important infrastructure projects such as large bridge crossings the operation of the bridges before, during and after the rehabilitation and strengthening is the key issue. This is due to the direct and indirect economic consequences of temporal and permanent traffic restrictions. Direct costs may be due to the loss of income such as tolls. Indirect costs may be due to loss of reputation.

Modern bridges are designed with special emphasis on minimising the imposition on the bridge operation due to maintenance and rehabilitation works. In this way structural joints, bearings and other utilities are designed so as to allow for easy access, maintenance and exchange. However, also more significant structural elements such as stay cables on stay cable bridges and hanger cables on suspension bridges are designed in such a way that they may be exchanged with a minimum impact on the normal operation of the bridge.

A significant number of bridges have not been designed in due consideration of the deterioration processes actually acting on the structures. Furthermore, the design of existing structures has often not taken into account the actually performed inspection and maintenance.

For such bridges there may be considerable restrictions on the allowable traffic associated with strengthening and repair activities. In these cases it is important to have access to an efficient planning tool to minimise the restrictions to traffic, and at the same time to ensure that the bridge is appropriately safe during and after the rehabilitation.

Structural reliability methods have proven their value through numerous applications over the last decades. Without giving references to the vast literature on applied reliability analysis it is here just emphasised that a fundamental strong point about reliability methods is their ability to support a consistent ranking of structures and load cases in terms of reliability. This ability is precisely what is required for the rational planning of rehabilitation.

By evaluating the reliability of the bridge corresponding to the condition of the bridge in the different phases of the rehabilitation and by taking into account the different practically possible traffic restrictions it is possible to identify the maximum allowable traffic on the bridge during the rehabilitation and at the same time to ensure an appropriate level of safety.

The present paper briefly describes a format for rehabilitation, introduced as the Rehabilitation Design Basis (RDB), which has been designed especially for this purpose. The benefits and limitations to the use of the RDB are illustrated and discussed through a recent application concerning the emergency rehabilitation of the Zárate-Brazo Largo bridges in Argentina.

2. FRAMEWORK FOR REHABILITATION DESIGN

2.1 General considerations

The design and assessment of ordinary structures under normal conditions is appropriately accommodated by most existing codes of practice and regulations. This is because most existing codes and regulations have been formulated and calibrated specifically to ensure that the most commonly built structures under normal conditions are both economic and sufficiently safe.

For the design and assessment of structures which are unique by e.g. proportion, concept, material or condition, codes and regulations, cannot be expected to yield structures which are appropriate in terms of economy and safety. For the design and assessment of such structures it is therefore common practice to formulate and calibrate a design basis which is specific for the considered structure.

Such a design basis can be understood as a specific code of practice for the design and assessment of this unique structure.

2.2 Basis for design of rehabilitation

Bridges subject to significant deterioration or subject to major redefinition of their operating conditions deviate significantly from normal structures. For this reason such structures may therefore be considered as unique. The existing codes and regulations may thus be inadequate for use as basis for verification of load carrying capacity and residual service life as well as for rehabilitation design.

In order to ensure that a bridge is efficiently rehabilitated to a condition where it may be accepted for safe continued use, and possibly even upgraded in terms of traffic loading or service life, it is necessary to formulate and calibrate a rehabilitation design basis specific for the considered type of bridge, the observed state of deterioration of the bridge, and the desired use of the bridge during and following the rehabilitation.

An appropriate safety format for the rehabilitation design basis is the Load and Resistance Factor Design (LRFD) format. By specifically assigning characteristic values and partial safety factors to the individual design variables in consistency with the available information, design values for the load and resistance variables may be derived for different rehabilitation phases and different load situations.

In general, the basis for the calibration of the characteristic values and the partial safety factors is the modern reliability theory, see e.g. Ditlevsen and Madsen [1]. This approach complies with the principles described in the background documents to the Eurocodes and the ISO codes, see e.g. [2] and [3], and is thus internationally accepted for calibration of codes.

2.3 Updating of the rehabilitation design basis

It is essential to update the design values for load and resistance variables as new information about these becomes available.

The rehabilitation design basis is therefore formulated so that it can be modified or updated in regard to additional information such as
- information regarding past loading
- assumptions regarding future loading
- results from material tests
- measurements of forces and displacements
- inspections in regard to structural deterioration
- strengthening and replacement
- traffic restrictions

In figure 1 the relation between typical rehabilitation actions, knowledge about the structure and use of the structure is illustrated. The grey toned

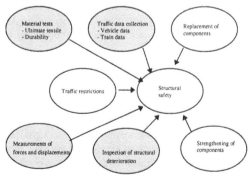

Figure 1   Illustration of the relation between structural safety, available knowledge and intended use of the bridge.

"knowledge collecting" actions are the activities which may continue throughout the rehabilitation until a sufficient basis for the final rehabilitation has been established.

3. PROCEDURE AND REQUIREMENTS FOR ESTABLISHING A REHABILITATION DESIGN BASIS

3.1 *Probabilistic load models*

As a first step, probabilistic models are established for the loading on the bridge. It is of course important that these models are reasonably precise, however, the most important issue is that the principal characteristics of the loads and the uncertainties associated with the prediction of the loads are captured in the models. This consequently implies that the load models shall be able to represent the actual loading in terms of any observable or measurable quantity of the loading. For roadway traffic load models such quantities are e.g. the number of trucks per time unit, the distribution of trucks on the different lanes, the relative frequency of trucks of different weights, etc.

3.2 *Probabilistic resistance models*

Secondly, a probabilistic model for the resistance of the components of the bridge is established. Again the most important issue is to capture the principal characteristics of the durability and ultimate load carrying capacity. As for the loads it is important that the models relate to observable quantities such as e.g. tests results from the laboratory regarding the ultimate tensile strength and the fatigue life of specimens taken from the bridge. However, also structural response characteristics to known excitation may be a valuable input to the resistance models.

3.3 *General model considerations*

Common for the probabilistic load and resistance models is that they shall be able to represent the uncertainties due to the differences between the models and the observations of the real world, i.e. the model uncertainty and the physical uncertainty. In addition the models shall be able to represent the uncertainty arising from lack of data, i.e. the statistical uncertainty.

As a basis for the probabilistic modelling of load and resistance characteristics it is recommended to take advantage of the Bayesian statistical framework, see e.g. Lindley [4]. Especially for the application of parametrical models being fitted to observation or test data, this framework has proven its usefulness. Furthermore, the Bayesian statistics provide a consistent framework for the combination of engineering judgement and information obtained through observations and tests.

3.4 *Calibration and requirements to reliability*

It is well known, but not always fully appreciated, that the reliability of a structure as estimated on the basis of a given set of probabilistic models for loads and resistances has only limited bearing to the actual reliability of the structure. For this reason it is not immediately possible to judge whether the estimated reliability is sufficiently high without first establishing a more formalised reference for comparison.

In principle two approaches are available to establish a reference of reliability. Common for both approaches is that the reference for comparison of reliability is established by consideration of an "optimal" structure. This structure is referred to as a "best practice" bridge design. The idea behind the "best practice" reference is that if the bridge considered has been designed according to the "best practice" then the reliability of the bridge is "optimal" according to agreed conventions for the target reliability. Examples of agreed conventions for the target reliability of structures may be found in e.g. NBK [5]. Typical values for the corresponding target annual failure probability are in the range of 10^{-6} to 10^{-7} depending on the type of structure and the characteristics of the considered failure mode.

Using the first approach, the probabilistic models are calibrated so that the reliability of the structure equals the target reliability. In the second approach the probabilistic models are not calibrated but the target reliability is determined as the reliability of the "best practice" bridge with the given probabilistic model..

The determination of the "best practice" design can be performed in different ways. The simplest approach is to use the existing codes of practice for design as a basis for the identification of "best practice" design. Alternatively the "best practice design" may be determined by consultation of a panel of recognised experts.

However, whereas the most practical approach would be a combination of the two above mentioned approaches the most rational approach is to establish the "best practice" design on the basis of the economic decision theory. This aspect is discussed in detail in Ditlevsen [6].

4.   DETERMINING THE REHABILITATION
SAFETY FORMAT

Having established and calibrated the probabilistic models for loads and resistance characteristics of the bridge the next step is to establish the rehabilitation design safety format. The necessary steps to do this are to:
1.   identify rehabilitation phases
2.   determine characteristic values for loads and re-sistance variables
3.   determine partial safety factors for load and re-sistance variables
4.   determine load combination factors

The first step is necessary in order to identify the structural configurations and corresponding optional load conditions during all intermediate phases of the rehabilitation as well as in the final rehabilitated state of the bridge. In this step it is important to carefully consider the possible practical traffic situations and corresponding traffic restriction on the bridge in the different phases. This will allow for a minimisation of traffic inconveniences resulting from the rehabilitation.

In the second step the characteristic values for the resistance and the load variables are determined. Reference to a common definition of characteristic values of resistances and loads may be found in e.g. Eurocodes 1.1, Basis for Design [7]. In the following a brief description of characteristic values of the most important variables for bridge rehabilitation is given.

The characteristic values for resistance variables are usually determined as a lower fractile value of the corresponding probability distribution function. Typically the 5% fractile value is used for most resistance characteristic values.

The characteristic value of variable loads is usually defined by the 98% percentile value of the probability distribution function for the annual maximum load effects. As variable loads, such as roadway and railway traffic loads, in reality are comprised of complicated sequences of axle loads in time and space, characteristic loads for these are normally defined in terms of Equivalent Uniformly Distributed Load (EUDL) patterns. The intensity of the EUDL is calibrated so that the resulting dominating load effect equals the 98% percentile value of the probability distribution function for the annual maximum load effects. This principle is illustrated in figure 2 and 3 for road and railway loads respectively.

It is important to notice that in principle different characteristic loads, i.e. EUDL's may be identified for

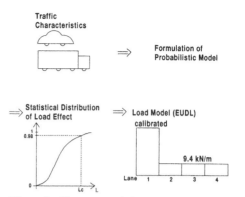

Figure 2   Illustration of the principle for calibration of characteristic roadway traffic loads.

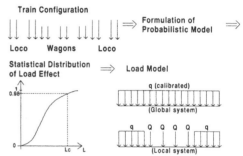

Figure 3   Illustration of the calibration of characteristic railway loads.

different bridges or different components of bridges even though the bridges and components are subject to the same traffic loading. This is due to the fact that the characteristic load is defined in terms of the statistical characteristics of the load effects rather than in terms of the loading itself. Bridges and components of bridges which filter the loading differently consequently have different characteristic loads. In design situations it is more practical to define the characteristic loading in correspondence with the largest load effect in the bridge. However, in rehabilitation design it may be more rational to consider bridge component specific characteristic loads.

Characteristic loads for permanent loads such as dead loads are typically defined in terms of their mean value.

The third step is to determine partial safety factors for all resistance and load variables. The principles for the determination of partial safety factors are described in e.g. JCSS [2] (Background document to Eurocodes 1991-1). The underlying principle for the determination of partial safety factors is that the safety factors shall reflect the

- importance of the individual variables in the most likely realisation of variables given failure
- uncertainty associated with the individual variables

The safety factors for resistance variables R, permanent loads P and variable loads Q may be determined by equations of the form.

Finally, in order to take into account load cases where several loads may act simultaneously, load combination factors are determined.

Load cases are defined in terms of a main load and one or more companion loads.

The load combination factors are multiplied on the companion loads in order to take into account the rareness of simultaneously acting loads.

The load combination factors may thus be evaluated in terms of the number of relative changes in load intensity of the combined loads, together with information regarding the variability of the loads.

5. THE ZÁRATE - BRAZO LARGO BRIDGES

5.1 *Presentation of the bridges and the problem*

The Zárate - Brazo Largo bridges are part of a major infrastructure project in Argentina providing a crossing of the national road No 12 over the two main branches of the Paraná river - Paraná de Las Palmas and Paraná Guazú - northwest of the river delta close to the town of Zárate.

The two bridges constitute two cable stayed steel girder bridges with main spans of 330 metres and a total of about 16 km concrete approach spans for railway and roadway traffic.

The bridges were constructed during the years 1972-1977, based on an alternative bid submitted by the Joint Venture Techint - Albano, and opened to roadway traffic in 1977 and railway traffic in 1978.

A photograph of one of the two bridges, the Guazú bridge, is shown in figure 4.

Stay cable 7C of the bridge across Paraná Guazú ruptured on 20 November 1996 in the morning. The stay cable failed close to the bottom socket which plunged into the river and was recovered by divers two days later.

The remaining part of the cable was removed from the upper anchorage and put on the bridge deck. It showed that almost all of the 121 wires Ø 7 mm failed about 200 mm away from the bottom socket with severe signs of corrosion and fatigue-like ruptures in the wires.

Figure 4 Photograph of the bridge crossing the Paraná Guazú river in Argentina.

5.2 *Immediate actions and approach to the problem*

As an emergency action, the bridges were closed to traffic on 25 November 1996, and the consortium Albano - DyCASA - Freyssinet (ADF) was entrusted by Dirección Nacional de Vialidad, DNV, the national highway authority of Argentina, with the immediate replacement of stay 7C, using the Freyssinet mono strand stay type.

The general philosophy adopted for handling the situation was to

- stabilise the safety of the bridges by
  - exchange and strengthening of stay cables
  - imposing restriction to the traffic
- investigate the cause(s) for deterioration
- prepare the bridges for final rehabilitation

It was decided to base the assessment and the rehabilitation on the framework of the Rehabilitation Design Basis described previously. This in order to ensure that the rehabilitation would utilise to the largest extent possible the available information regarding the condition of the bridge and be able to relate the reliability of the bridges during the rehabilitation with the traffic condition in general and possible traffic restrictions in particular.

In the following the most essential points in order to derive probabilistic models for the strength and durability of the stay cables are described. Secondly the probabilistic modelling of the bridge loading is addressed. Based on the probabilistic load models the characteristic loads to be used for the design of strengthening and exchange of stay cables are derived in accordance with different load conditions and restrictions considered during the rehabilitation.

5.3 Models for ultimate strength and durability

The stay cables used in the design of the bridges are parallel wire cables with HiAm sockets.

The characteristic strength of the wires and the corresponding partial safety factors for the stay cables have been derived on the basis of a probabilistic model of the time varying strength of parallel wire cables subject to fatigue deterioration. The model incorporates all available information regarding the material characteristics of the wires as obtained through testing under ultimate tensile as well as fatigue loading.

Results of intensive ultrasonic testing of the integrity of the wires together with the results of a large number of ultimate tension and fatigue test performed on wires in the laboratory constitute the most essential data to the evaluation.

The model for the safety factor of the stay strength allows for a differentiation in safety factor in accordance with the intended service life of the considered stay. Furthermore, the safety factors may be derived for the stays individually, taking specifically into account the knowledge about the damage condition and the loading on the individual stays.

Using the model it is possible to derive safety factors corresponding to intermediate periods for the bridge (during rehabilitation) which are lower compared to the situation where a normal service life is considered. Furthermore the model also yields safety factors to be used for the design of stay cable replacements.

The computation of the residual life is based on a method proposed by Rackwitz and Faber [8]. The first step in the computation of the residual life is the estimation of the distribution of the fatigue life of a single wire.

The number of cycles, t, to failure of a given wire subject to a stress range, Δs, is given by a Weibull distribution,

$$F_T(t, \Delta s) = 1 - \exp\left(-\left[\frac{\Delta s}{r_c}\right]^{\alpha}\left[\frac{t}{K}\right]^{\frac{\alpha}{m}}\right) \quad (1)$$

where α, m, K, $r_c = \left(cA_0 n\right)^{\frac{1}{\alpha}}$ and c are unknown parameters, n is the number of sections of the test specimen with different properties and A_0 is the cross sectional area of the wire.

In order to utilise the fatigue life model for individual wires in the assessment of the fatigue life of parallel wire bundles some assumptions are made. First of all, the time to failure of the individual intact wires given a certain level of mean stress and stress range is assumed to be identically and independently distributed. Also it is assumed that the initial load of the cable is small enough to cause no static wire failure.

Failure of a parallel wire bundle due to fatigue occurs as a sequence of failures of individual wires. One wire breaks after the other due to accumulated fatigue damage. The wire with the shortest time to failure, fails first in each state of the system. Immediately after failure of each of the individual wires a load redistribution takes place. It is assumed that this load redistribution takes place without dynamic effects.

The residual static strength of the individual wires is assumed not to be influenced by the accumulated fatigue damage. Finally, it is assumed that there is no dependency between the static strength of a given wire and the fatigue resistance of the wire.

It may be shown that the number of fatigue load cycles t until a or the proportion y of the cross section has failed can be written as

$$t(y) = \frac{r_c^m K_0}{\beta} \int_0^d \left[\frac{\frac{\Delta s}{1-z}}{\left[1 - \frac{m_s}{(1-z)m_z}\right]^{\frac{\gamma}{m}}}\right]^{-m} \left[\ln\left[\frac{1}{1-z}\right]\right]^{\frac{1}{\beta}-1} \times$$

$$\left[\frac{1}{1-z}\right] dz \quad (2)$$

with $\gamma = 0.5$.

The effect of corrosion is taken into account by determining a set of material parameters specifically for corroded wires. Furthermore, corrosion implies that the length of wire for which the material parameters can be assumed to be constant, becomes small. Therefore, the parameter r_c which depends on n (the number of parts with different material parameters) depends on whether the wire is corroded.

Having estimated the number of wires remaining in the stay cable as a function of time it remains to determine the corresponding strength.

The distribution function $F_Z(z)$ for the strength z of a wire of length L may appropriately be given by a Weibull distribution with parameters λ, u and k as

$$F_Z(z) = 1 - \exp\left[-\lambda\left(\frac{z}{u}\right)^k\right] \quad (3)$$

$$E[Z] = u\,\lambda^{-1/k}\,\Gamma\left(1 + 1/k\right) \quad (4)$$

$$V[Z] = u^2\lambda^{-2/k}\left[\Gamma(1 + 2/k) - \Gamma^2(1 + 1/k)\right] \quad (5)$$

where the scale factor λ is given by

$$\lambda = \frac{L}{l L_0} \qquad (6)$$

and where L_0 is the length of the reference (test) wire specimen and $l \times L_0$ is the correlation length of the material parameters and/or the defects in the wire.

The ultimate tensile strength of individual wires may be characterised by the mean value μ and the standard deviation σ of the ultimate rupture stress determined experimentally on wire specimens of reduced length L.

For new and undamaged wires it may be assumed that the correlation length is in the same order of magnitude or even larger than the length of the considered wire e.g. $\sim$ 1000 metres. For old or damaged wires the correlation length may be reduced to a length in the order of the diameter of the wire e.g. $\sim$ 7 mm.

The actual correlation length $l \times L_0$ together with the parameters k and u may be estimated from experimental ultimate capacity tests by the Maximum Likelihood Method. Having observed from ne experiments the ultimate capacities $\underline{x}$ the parameters k, u and l may be estimated from

$$\max_{u,k,l}(\mathbf{L}(u,k,l)) \qquad (7)$$

where the likelihood function $\mathbf{L}(u,k,l)$ is given by

$$\mathbf{L}(u,k,l) = \prod_{i=1}^{ne} F_{\underline{X}}(\underline{x}_i, u, k, l) \qquad (8)$$

$$F_{\underline{X}}(\underline{x}_i, u, k, l) = 1 - \exp\left[-\frac{1}{l}(\frac{x_i}{u})^k \right] \qquad (9)$$

As is seen from equation (3) l and u are represented in the problem only in terms of a product between l and u^k. This functional relationship poses a problem when the maximisation according to equation (6) is performed as there is no unique solution for l and u but rather for $l \times u^k$. However, as the correlation length (the product between l and L_0) may be assumed to be a constant for test specimen of different length L_0 this problem can be overcome by testing wire specimens of different length.

The strength of wire bundles depends on a number of factors such as the number of wires, the length of the wires, the correlation length for failures in the wires.

The strength of bundles of parallel wires may be assessed through parallel systems considerations using the Daniels model.

The strength of a parallel system with n components may, if n is large enough (n > 150), be shown to be normal distributed with mean value

$$E_n = n X_0 (1 - F_Z(X_0)) + C_n \qquad (10)$$

and standard deviation

$$D_n = X_0 \left[F_Z(X_0)(1 - F_Z(X_0)) \right]^{1/2} \qquad (11)$$

where C_n may be assessed from

$$C_n = 0.966 n^{1/3} a \qquad (12)$$

and

$$a^3 = \frac{f_Z(X_0) X_0^4}{(2 f_Z(X_0) + X_0 f_Z^j(X_0))} \qquad (13)$$

$f_Z(X_0)$ is the density function for the wire strength. X_0 which may be determined from

$$X_0 = \left[l \frac{L_0}{Lk} \right]^{1/k} u \qquad (14)$$

C_n may be considered as a correction term to the asymptotic solution (which is valid for large n) in cases where n is below say 150.

Even for moderate number of wires (n > 100) the strength of a parallel wire bundle can be considered to be deterministic, which may also bee seen by application of the expression for the standard deviation given above. This property implies that the uncertainties influencing the failure probability for static failure modes of wire bundles may be attributed to the loading.

5.4 *Utilisation of inspections and tests*

To assess the remaining cross-sectional area of the individual stay cables an extensive ultra sonic inspection programme has been conducted by the DMT, Germany. By inspecting each wire from the wire end at the bottom and top sockets, see figure 5, it was possible to obtain indications of wire ruptures. An example of the representation of inspection results is given in figure 6.

In order to fully utilise the inspection results in the framework of structural reliability analysis it is necessary to determine the reliability of the UT-inspection method.

Figure 5 Photograph showing the UT-inspection crew at work.

The reliability of the method was assessed on the basis of a series of "blind" inspection trials performed on two dismantled stays in the laboratory. The trials were performed by different inspectors who were also active in the UT-inspections on the bridges. The inspection trials indicate that the probability of detecting a broken wire, the POD, by the in situ inspection can be modelled by a β-distribution with mean of about 0.90 and standard deviation 0.03.

Figure 6   UT-inspection results. Dark spots indicate wire ruptures/damages

Using this model for the probability of detecting a broken wire the cross-sectional areas of the cables have been estimated by a Bayesian analysis. The results indicate that there is only little uncertainty related to the remaining cross-sectional areas. In general the coefficient of variation of the cross-sectional area of a given cable is less than 2.5 %. This indicates that the information gained from the UT-inspections lead to an accurate prediction of the remaining cross-sectional areas.

5.5 Safety factors for stay cables

Based on the model for the static strength of the stay cables, the model of the deterioration of the stay cables and the results from UT-inspections, the static strength of the stay cables have been determined as a function of time. On the basis of this the partial safety factors for the stay cables have been determined with the requirement that the safety of the stay cables at the end of the considered period is in accordance with normal safety requirements to structural components of the same importance.

5.6 Roadway traffic loads

Probabilistic models have been formulated for the roadway traffic load on the basis of the models originally developed for the design of the Great Belt East

Bridge, see e.g. Ditlevsen [9] .

The load model has been calibrated to the extent possible on the basis of traffic observations from the Zaraté-Brazo Largo bridges. In some cases it has been necessary to supplement these observations with experience data from Denmark and Europe in general.

Information from the bridge regarding the number of trucks of different weight has been collected as indicated in figure 7.

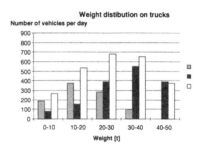

Figure 7   Distribution of trucks (category 4,5 and 6)

Using the probabilistic load models together with influence lines for a number of selected load effects (including stay forces, road girder load effect and pylon load effect), the statistical distribution of the maximum load effect during one year has been determined. By defining the characteristic load as the 98 percentile of this distribution, an equivalent uniformly distributed load (EUDL) has been derived which yields the same load effect as the characteristic load.

The input to the roadway load model is shown in table 1.

With the input shown in table 1 the EUDL for the roadway traffic load has been determined as shown in table 2. The axle loads have been taken from the Eurocode [10] without further calibration.

The EUDL load pattern given in table 2 has been derived on the basis of a EUDL for all four lanes of 16.2 kN/m by setting the least loaded lanes to the same intensity given in the Eurocode ENV 1991-3 [10] (Traffic Loads on Bridges) and by adjusting the intensity of the heavy loaded lane such that the overall load effect is maintained.

The results of a sensitivity analysis regarding the influence of the different input to the roadway load EUDL is shown in figure 8.

From figure 8, it is seen that by far the most important parameters are the weight of the trucks, the length of the cars (keeping distance between the

Table 1 Input parameters to roadway traffic load model.

Variable	Input
Number of lanes	4
Length of influence line L	550 m
Speed V in m/hour	75000 m/hour
Length of car	6 m
Weight of car	10 kN
Mean weight of truck	258.3 kN
Standard deviation of truck weight	119.8 kN
Fraction of truck among vehicles	0.30
Distribution of trucks on lane 1 in congested traffic	0.90
Distribution of trucks on lane 2 in congested traffic	0.10
Distribution of trucks on lane 4 in free traffic	0.95
Distribution of trucks on lane 3 in free traffic	0.05
Distribution of cars on lane 4 in free traffic	0.80
Distribution of cars on lane 3 in free traffic	0.20
Platoon length parameter	0.67
Area of influence function lane 1	23.046
Area of influence function lane 2	18.272
Area of influence function lane 3	13.784
Area of influence function lane 4	10.056
Area of squared influence function lane 1	5.2244
Area of squared influence function lane 2	3.6650
Area of squared influence function lane 3	2.2398
Area of squared influence function lane 4	1.2854
Intensity of congested situations per year	6.1/year
Intensity of trucks in free traffic (one direction)	300/hour
Intensity of cars in free traffic (one direction)	700/hour
Mean duration of congested situations	1.8 hour
Reference time in years	0.25

Table 2 EUDL and axle loads for the roadway traffic loads.

Location	Uniformly Distributed Load [kN/m]	Axle Loads according to EC [kN]
Lane 1	28.932	2 x 300
Lane 2	9.375	2 x 200
Lane 3	9.375	2 x 100
Lane 4	9.375	0

trucks) and the area of influence for the different loaded roadway lanes.

5.7 Railway traffic loads

The railway load model has been formulated on the basis of the model developed for the design of the Great Belt West bridge Madsen [11].

The configuration of the train considered for the determination of the characteristic railway load is:
Two locomotives in front:
    total weight (one locomotive) = 94 t

total length (one locomotive) = 15.5 m
uniformly distributed load = 6.06 t/m
axle weight = 15.66 t
Followed by 45 wagons
    total weight (one wagon) = 80 t
    total length (one wagon) = 12 - 14 m
    uniformly distributed load = 6.67 t/m
    axle weight = 20.0 t
One locomotives at the end
    total weight (one locomotive) = 82 t
    total length (one locomotive) = 15.3 m
    uniformly distributed load = 5.36 t/m
    axle weight = 13.66 t

The weight and length of the locomotives are modelled deterministically with weight and length data as given above. The weight of the wagons is modelled probabilistically. The assumptions regarding the modelling are as follows:
- Maximum weight of a wagon is 80 t.
- Wagon length is equal to 12 m.
- The wagon weight is normally distributed.
- The coefficient of variation of the wagon weight is 0.03.
- The maximum wagon weight corresponds to the mean value plus two times the standard deviation.
- The wagon weight is as a result of the above items modelled with a mean value equal to 62.9 kN/m and a standard deviation equal to 1.9 kN/m.
- The number of train passages is 2 per day.

The evaluation of the characteristic railway load is shown in table 3 for different bridge components where the most unfavourable train length has been used.

Table 3 Characteristic railway load EUDL.

Stay	EUDL
1C	68.8 kN/m
3C	65.7 kN/m
4C	69.4 kN/m
7C	66.5 kN/m
12C	70.2 kN/m
Road girder	65.4 kN/m
Rail girder	63.1 kN/m

It is seen that the largest EUDL equal to 70.2 kN/m is achieved for stay cable 12C.

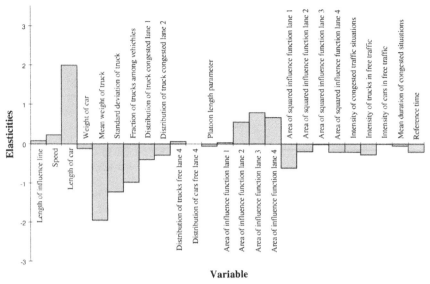

Figure 8   Results of sensitivity analysis of the EUDL

6. IMPACT OF TRAFFIC RESTRICTIONS

6.1 *Roadway loading*

During the emergency rehabilitation period different traffic situations have been evaluated in order to minimise the considerable inconvenience of traffic restrictions and to access alternative characteristic road loads.

The characteristic roadway load (EUDL) has been evaluated for the following restrictions on the 4 lanes:

1. minimum 100 m between the trucks in lane 1 and 2 but no restriction on the traffic in lane 3 and 4
2. only small trucks (with a mean weight of 5 tons) allowed on the bridge in all 4 lanes (including 7 tons busses)
3. no traffic in lane 1 but normal traffic without restrictions in lane 2, 3 and 4
4. normal traffic in all 4 lanes but the trucks are only allowed to pass in lane 1 and 4.

In the following the resulting EUDL for the different traffic situations are given. The evaluations are carried out considering the load effect in the road girder for the load case with congested traffic in one direction and free traffic in the other direction.

On this basis the the impact of the different traffic restrictions on the EUDL is.

1. Assuming that only 4 trucks are allowed in lane 1 and 2 at the same time (minimum 100 m be-

tween each truck) and no restrictions on the traffic in lane 3 and 4 yields an EUDL equal to 10.3 kN/m.

2. If only small trucks with a mean weight of 5 tons are allowed on the bridge (including busses weighing maximum 7 tons) the resulting EUDL is shown in table 5 for varying platoon lengths.

Table 5   Parameter study for the platoon length when small trucks are allowed

Platoon length (vans)	EUDL [kN/m]
1	2.6
2	3.0
3	3.3

3. If lane 1 is closed for traffic but no restrictions are imposed on the traffic in lane 2, 3 and 4 the EUDL equals 11.4 kN/m.

It is reasonable to believe that the intensity of congested situations per year will increase if lane 1 is closed. Therefore a parameter study for this quantity has been performed and the results are given in table 6.

It is seen that the number of congested situations per year has only minor importance for the roadway EUDL.

4. If normal traffic i.e. no weight restrictions in all

Table 6 — Lane 1 closed but free traffic allowed in lane 2, 3 and 4.

Table 6   Lane 1 closed but free traffic allowed in lane 2, 3 and 4.

Intensity of congested situations (per year)	EUDL [kN/m] in all 4 lanes
6.1	11.4
12.2	11.8
18.3	12.0

4 lanes is allowed but the trucks only are allowed to pass in lane 1 and 4 the resulting EUDL is 13.8 kN/m.

6.2 Railway loading

For the railway traffic on the single track three different load situations are considered corresponding to three different categories of train transports, see table 7, where also the corresponding expected values and coefficient of variations are shown. The wagon weights are assumed to be normally distributed.

Table 7   Categories of wagons and loads.

	loads of wagons	expected value	coefficient of variation
heavy train	55-66 kN/m	60.5 kN/m	0.05
medium train	44-55 kN/m	49.5 kN/m	0.075
light train	33-44 kN/m	38.5 kN/m	0.10

It is assumed that the intensity of each of the three categories of trains is ν=730/year.

By evaluating the 98% of the distribution of the annual maximum load effect the EUDL's for the considered train categories shown in table 8 are derived.

Table 8   Railway EUDL for different bridge components.

	heavy train	medium train	light train
	EUDL [kN/m]	EUDL [kN/m]	EUDL [kN/m]
Road Girder	69.1	62.7	54.7
1 C	72.2	64.2	54.1
3 C	69.4	63.0	54.9
4 C	71.6	64.0	54.6
7 C	69.5	62.8	54.4
12 C	72.2	64.2	54.1

It is seen that the maximum characteristic EUDL rail loads are

Category	EUDL
heavy train :	72.2 kN/m
medium train :	64.2 kN/m
light train :	54.9 kN/m

The characteristic railway loads have been utilised during the rehabilitation for specifying the heaviest train class allowed on the bridges in a given rehabilitation phase.

6.3 Safety factors for road and railway loading

Design values for the loads to be used for the rehabilitation design have been derived directly on the basis of the characteristic values derived in the previous section together with the formulas for the partial safety factors and the load combination factors from e.g. JCSS [2].

As the safety factors depend on the variance of the distribution of the load effects it is important to note that changes in the traffic situations, e.g. by introduction of restrictions not only have impact on the characteristic loads but also on the safety factors and load combination factors.

7. CONCLUSIONS

Based on the theoretical framework of structural reliability theory a Rehabilitation Design Basis (RDB) has been formulated for the rehabilitation of bridges subject to deterioration, damages and/or changes in operation conditions.

The RDB allows for a reassessment of structures taking specifically into account any knowledge about the condition and the loading on the structure.

Therefore, the RDB is specifically well suited as a rational basis for evaluating the consequences of traffic restrictions on structural safety.

The application of the RDB is illustrated with basis on a recent experience from the emergency rehabilitation of the Zaráte-Brazo Largo bridges in Argentina.

8. ACKNOWLEDGEMENTS

The authors are thankful for the permission from Dirección Nacional de Vialidad to publish data from the ongoing rehabilitation of the Zaráte-Brazo Largo Bridges.

9. REFERENCES

[1]   Ditlevsen, O. and Madsen, H.O., Structural Reliability Methods, Wiley, Chichester, 1996.

[2]   Joint Committee on Structural Safety, Background Documentation, Eurocode 1 (ENV 1991) Part 1 : Basis of Design.

[3]   ISO/DIS 2394 General principles on reliability for structures, Revision of the first edition (ISO 2394:1986), 27-11-1996.

[4]   D.V. Lindley, *Probability and Statistics*, Cambridge University Press, Cambridge 1965.

[5]   Nordisk Komite for Bygningsbestemmelser, NKB-skrift 55, Sikkerhedsbestemmelser for bærende konstruktioner. NKB, 1987, in Danish.

[6]   Ditlevsen, O., Risk Acceptance Criteria in the Light of Decision Analysis, 22nd WEGEMT Graduate School, Accidental Loadings on Marine Structures: Risk and Response, Technical University of Denmark,24th-29th April, 1995.

[7]   Eurocode 1, ENV 1991-1: Basis of Design and Actions on Structures, Oct. 1993.

[8]   R. Rackwitz and M.H. Faber, *Reliability of Parallel Wire Cable under Fatigue*, in: Proceedings to the ICASP6 Conference, Mexico, 1991, Vol. 1, pp. 166-175.

[9]   H.O. Madsen and O. Ditlevsen, *Stochastic Traffic Load Modelling for the Eastern Bridge*, A/S Storebæltsforbindelsen, Review Board 3, May 1990.

[10] Eurocode 1.3, ENV 1991-3: Basis of Design and Actions on Structures, *Traffic Loads on Bridges*, June, 1994.

[11] Madsen, H.O. *Load Combinations*, AS Storebæltsforbindelsen, 1989.

Operation and Maintenance of Large Infrastructure Projects, Vincentsen & Jensen (eds)
© 1998 Taylor & Francis, ISBN 90 5410 963 7

The UK experience of design, build, finance and operate highway projects

N.W. Roden
Private Finance Team, Highways Agency, London, UK

ABSTRACT: In the United Kingdom, the Private Finance Initiative (PFI) is the procurement method of choice for many projects and services, enabling government to become a buyer of services on behalf of the public rather than a direct provider of those services. The PFI is delivering value for money solutions across a range of capital-intensive public services, but the greatest success has been in the transport sector. This paper outlines how the British Government's Highways Agency has restructured and commercialised the management and maintenance of the strategic road network as part of this change in focus of the public sector, and has developed its particular version of BOT - Design, Build, Finance and Operate (DBFO) projects - that are delivering new and improved road construction and maintenance to provide better services to users of the country's strategic road network.

1. INTRODUCTION

In today's world, an effective transport system is not an optional extra - it is a prerequisite of modern life. Many essential public services depend upon it, as does the success of industry and commerce. It is not surprising therefore, that in the United Kingdom as elsewhere, it is in the field of transport that progress in attracting private sector investment has been greatest.

This paper outlines the background to commercialisation policies in the UK, particularly in the area of transport. Complementary reforms were introduced in public administration to improve the efficiency of the public sector and the paper examines these management reforms designed to create better accountability and management in the efficient execution of government policy. It was these policies that led to the creation of the Highways Agency which is delivering services more efficiently and effectively, and becoming increasingly customer orientated as its role develops from being an organisation concerned primarily with building highway infrastructure to one of focusing on the operation of the network. Finally, the paper describes how, in the absence of legislation permitting the introduction of user paid tolls on the existing network, the Highways Agency has developed its particular version of BOT projects - Design, Build, Finance and Operate (DBFO) projects that deliver new and improved road construction and maintenance.

2. COMMERCIALISATION OF TRANSPORT SERVICES IN THE UNITED KINGDOM

The challenge facing governments throughout the world, is to deliver their objectives in a way which makes best use of all the resources at their disposal, both pubic and private. The public has a right to expect quality services delivered cost effectively and the means by which this is achieved is less important than the result. The private sector has strengths and skills which are not found in the public sector and by harnessing those skills, services can be delivered more effectively and efficiently. For example, the private sector knows how to act commercially because it understands the market place and the need for competitiveness. It knows that, to be successful, it must not only respond to the needs of its customers, but strive constantly to improve its services. The private sector is also often better placed to manage many of the risks traditionally borne by Government and it has access to new and more flexible sources of capital.

In line with this philosophy, the role of the state in the UK economy was redefined under the Conservative administrations of the 1980s and early 1990s, changing from that of a direct provider of services to one of enabler and regulator for the private sector provision of services. This was a key area of Government policy which led to significant changes in the UK transport and utilities markets. It is well illustrated in the extensive measures that were taken to liberalise transport, remove barriers to entry, and reduce regulation. Public services delivered by Government and for which the user paid directly, such as energy and telecommunications utilities, were privatised. Similarly, airports and airlines, many ports and, finally, railways were privatised.

To improve the efficiency of the public sector where the Government remains the direct provider or purchaser of services, other related reforms were introduced. The underlying principle of these changes was to introduce the best private sector practices into the business of public administration. These changes encompassed improving accountability and management in the delivery of central and local government services through the creation of Executive Agencies; requiring support services to be tested against private sector competition through market testing and compulsory competitive tendering; and introducing commercial accounting and reporting practices into the public sector.

In summary, whether through privatisation, contracting out, or the provision of services and infrastructure that remained within the public sector, Conservative administrations believed that the greater involvement of the private sector provided opportunities to ensure improved quality, faster results, and better value for money for the taxpayer.

The indications from the new Labour Government are that many of the significant structural changes in the public sector over the life of the previous administrations will be consolidated and developed rather than overturned. This is particularly so in the case of the Private Finance Initiative (PFI) where the newly elected Labour Government has taken steps to review and invigorate the initiative.

3. CREATING THE HIGHWAYS AGENCY

The English trunk road network for which the Highways Agency is now responsible was established over 60 years ago, although many of the roads in the network were first built long before that and some still follow the routes first laid out by the Romans some 2000 years ago. However, the recent history stems from the early 1930s and the growth of motorised road transport in the UK. At that time there was a growing concern that the tradition of local authorities looking after all roads could result in inconsistencies in approach, frustrating the development of facilities for national, long-distance, strategic traffic. Following a review it was decided that the most important routes should be designated as "trunk roads" and that central government, rather than local government, should be responsible for developing and maintaining this network to cater for "through traffic" in the national interest.

For the last 60 years Government Ministers (currently the Secretary of State for the Environment, Transport and the Regions) have been directly responsible for the policies applying to trunk roads which, since 1959, have also included motorways. They are also responsible for determining the resources to be spent on the network and the priorities to be given to spending and major new highway schemes. Initially, the funding for road improvements and maintenance was provided by the revenue obtained from vehicle licence fees but, by the 1940s, this revenue was subsumed within general taxation. Today, all revenue from UK motorists either in the form of vehicle licence fees, driver licence fees, or fuel duty, is treated as general taxation and the annual expenditure on trunk roads is decided as part of the overall spending decisions of the Government.

Prior to 1994, the motorway and trunk road network had been managed directly by the Road and Vehicle Safety Command of the former Department of Transport. This part of the Department was also responsible for road and vehicle safety issues, and for wider policy, regulatory, advisory, and grant allocation functions relating to the whole road network, including that part of the network managed by local highway authorities.

However, as outlined above, since 1988 Government Departments have largely been reorganised in a way which focuses on the job to be done to enhance the effective delivery of policies and services. Under this initiative, known as "The Next Steps Programme", each of the executive functions of central government was examined to consider how it should be provided in the future and, if it were to remain within government, the best framework for its management. This has resulted in responsibility for the delivery of many Government services being transferred to Executive Agencies operating with a certain amount of autonomy and greater management freedom.

Despite the range and diversity of their functions, Executive Agencies share common features. Each agency has a clearly defined task, or range of tasks, which are set out in its published framework document, with key performance targets covering financial performance, efficiency, and service to the customer set by Ministers and announced in Parliament. Most targets have become progressively more demanding year on year, reflecting the need to ensure that limited public resources are used with the maximum of efficiency.

Each agency has a Chief Executive who is directly accountable to Ministers and with personal responsibility for the success of the agency in meeting its targets. The majority of chief executives have been appointed through open competition and more than half of those who have been appointed in this way have come from outside the Civil Service. Staff in Executive Agencies remain civil servants, accountable to Ministers and governed by the same strict rules of conduct as those working in departmental headquarters.

The need to enhance the effective delivery of services which these Executive Agencies are designed to achieve was reinforced by the former administration's "Citizen's Charter" programme. Agency framework documents and business plans emphasise customer service and these documents set out the service which the public can expect, together with arrangements for consultation and redress. All agencies are expected to consult users of their services about their priorities for improving them within what can be afforded, and many agencies have shown significant improvements in the quality and responsiveness of its service to customers.

An organisational review of the former Department of Transport's Road and Vehicle Safety in 1992 concluded that the main opportunities for transferring functions to an agency or agencies lay with the management of the motorway and trunk road network. That area of work represented a reasonably discrete business and within the business, the client side functions of strategic planning and control that needed to stay within the central Department were already reasonably well separated from the delivery of the agreed work programmes.

A number of alternative options were considered, but the then Conservative Government concluded that, establishing a single Executive Agency responsible for managing and maintaining the motorway and trunk road network and delivering a set programme of improvement works, was the option that would deliver the most significant benefits. As a result, the Highways Agency was established on 1 April 1994.

In general terms, the central Department retains the capacity to advise Ministers on general roads policy, the size and shape of the Agency's programmes, decisions about schemes to be included in the Government's road improvement programme, and the priority with which schemes should be delivered. The Agency is responsible for delivering an efficient, reliable, safe, and environmentally acceptable motorway and trunk road network. Its core functions are to manage and maintain the network, and to deliver improvements to the network within the strategic policy framework and the financial resources set by Government Ministers.

4. TRUNK ROAD AGENCY MANAGEMENT ARRANGEMENTS

When the Highways Agency was established in 1994, much of the day-to-day work of managing and maintaining the 10,400km motorway and trunk road network was undertaken by agents acting on behalf of the Highways Agency. Of the 91 of these agency agreements, 85 were with local highway authorities and the remainder with private sector consultants. These agency agreements covered the management of routine, winter and capital maintenance, and a wide variety of other duties ranging from inspections of the network to accident investigation and data provision. They also covered identification, design and supervision of capital maintenance works and minor road improvement schemes.

These standard agency agreements were first introduced in the 1940s and, at the time of the Highways Agency's launch, Ministers set the

Agency the task of keeping the arrangements with its managing and maintaining agents under review with a presumption in favour of increasing opportunities for the private sector to participate in the work. This is not to say that the Agency or Government Ministers were dissatisfied with the existing service provided by the local highway authority agents. Rather, there was a desire to benefit from the opportunities that competition might bring and to encourage a market of potential suppliers for managing and maintaining the network encompassing both the public and private sectors. The motorway and trunk road network is having to handle greater traffic volumes than ever before. It therefore made sense to maximise the Agency's options for ensuring that the network was maintained to an appropriate standard and for achieving a more consistent approach on a route basis.

In addition, there were a number of other initiatives which suggested a need for change such as the planned reorganisation of local government to create new unitary authorities and the introduction of compulsory competitive tendering requiring agent authorities to market test a proportion of their own professional services.

In April 1995, the Highways Agency launched a consultation exercise seeking views about its proposed changes to the management arrangements with agents. The objectives of the proposals were to:

1. continue to meet the statutory requirements for ensuring that the motorway and trunk road network was maintained to a satisfactory standard;

2. introduce more competition into the provision of services for managing and maintaining the network through the development of a private operating sector but to do so in a structured way which took proper account of the needs of trunk road users and delivered value for money; and

3. achieve an arrangement of management agencies that could deal efficiently with the demands on the network and were responsive to change.

In essence, the proposals were to replace the separate agency agreements with local authorities and consultants with a single agreement against which prospective public and private sector agents could compete on equal terms, and to obtain an arrangement of agencies based on strategic considerations. The outcome of the process has been to reduce the number of agency areas from 91 to 24, details of which were announced at the end of the consultation process in March 1996. These new agency management arrangements are being introduced on a phased basis over three years and following a competitive bidding process, the first contracts were awarded in 1996 and took effect in April 1997. By April 1999, contracts will have been tendered and awarded for all 24 agency areas.

The main features of the new motorway and trunk road management agency agreements are to introduce greater competition into the market and encourage the development of a private sector roads operating industry. This has required changes in the law to facilitate competition on an equal basis between local authorities and the private sector, and to enable certain statutory functions, previously only permitted to be undertaken by highway authorities, to be delegated to private sector agents. The new agency agreements are for a fixed term of three years with possible extensions for a further two years. Payment mechanisms have changed significantly under the new contracts. Payment is now based largely on an annual lump sum to cover the varied and wide-ranging duties of the managing agents and payment is made, in arrears, on presentation of a monthly invoice. The new arrangements require the introduction of a much more effective business planning and monitoring regime.

Under these new agency arrangements, the main duties of the managing agents are to design and supervise all road and structural maintenance activities, and network enhancement projects with a capital works cost of less than £1 million. They are also responsible for managing all routine maintenance and defects repair work, as well as general duties to ensure that the network is maintained in a satisfactory manner and that the Secretary of State's statutory obligations are discharged. Managing these activities on the basis of longer routes and bigger geographical areas allows for a more coordinated approach along substantial lengths of the network, ultimately helping the development and implementation of route strategies and performance criteria.

5. THE PRIVATE FINANCE INITIATIVE

Increasingly, the British Government is looking to partnerships with the private sector under the Private Finance Initiative (PFI) to deliver public sector projects and services of a higher quality more cost effectively.

Public/private partnerships are all about negotiating deals that are good for both sectors. The private sector wants to earn a return on its ability to invest and perform while the public sector wants contracts where incentives exist for the private sector supplier to deliver services on time and to specified standards. In that, there is a commonality of interest between the public sector procurer and the private sector financiers whose return on investment depends on these services being delivered to those standards.

As with privatisation and contracting out, the policy was introduced as part of the shift in the focus of the British public sector, away from being the direct owner and operator of assets to a purchaser of services from the private sector on behalf of the public. This transformation of the public sector away from buying asset based projects to a service orientated activity is continuing under the present administration. This means that the Government may no longer construct roads, but purchase miles of properly maintained and operated highway. It may no longer build prisons, but buy custodial services. Neither does it always buy computers and software, preferring to pay for managed IT services. The PFI differs from privatisation because the public sector retains a substantial role, either as the main purchaser of the services provided, or as an enabler of the project and guardian of the users' interests. It differs from contracting out in that the private sector provides the capital asset as well as the service.

One of the first acts of the new Labour administration was to commission a review of the PFI in line with their election manifesto commitments to develop public-private partnerships in sectors such as education and housing. The outcome of the review was that the Government took steps to introduce changes to the management of projects with the objective of revitalising the initiative. The PFI is not, therefore, likely to be replaced or substantially changed, but will remain one of the primary tools for providing many capital intensive government services.

The first PFI projects, particularly in sectors such as transport, prisons and information technology, have demonstrated the viability of the initiative in these key areas of central government procurement. The diversity of projects indicates that, for central government at least, there is no major impediment to its application across large areas of its activities. The models for central government PFI projects, such as those for DBFO highway projects, are also being used increasingly by local government to deliver the services for which they are responsible.

The main focus of PFI activity to date has been in the area of services sold to the public sector and, in particular, projects where the public sector purchases services from the private sector which is responsible for the "up-front" investment in the capital assets to support those services. The public sector client pays only on delivery of the services to the specified quality standards. Typically, the private sector, often acting in consortia, aims to reap synergies across design, build, finance and operation. This restructuring of procurement is mutually beneficial for users of public services, taxpayers, and companies seeking profitable new business opportunities. Despite the additional private sector borrowing costs and the necessity for the private sector service provider to make a profit, combining the private sector's innovation and management skills across the design, construction and operational phases of these projects, generates significant performance improvements and efficiency savings delivering improved value for money for the public sector client. DBFO road projects are an example of the successful application of these principles in the delivery of road services.

However, there are also a number of financially free-standing projects in the UK where the private sector supplier designs, builds, finances and operates an asset, recovering the costs entirely through direct charges on the users of the asset rather than through payments from the public sector. The concept of these Build, Operate, Transfer (BOT) or Build, Own, Operate, Transfer (BOOT) projects is well known and understood. The public sector's involvement is limited largely to enabling the project to proceed through assistance with planning and other statutory procedures. Apart from this, there is no public sector contribution or acceptance of risk. The primary examples of this type of project in the UK are in the transport sector where the Dartford Crossing of the Thames to the east of London and the Second Severn Crossing between England and Wales have been provided in this way. In addition, the Birmingham Northern Relief Road in the West Midlands which will be the first overland tolled motorway in the UK, is also being taken forward by the private sector as a financially free-standing project.

6. THE DEVELOPMENT OF DBFO CONTRACTS IN THE UNITED KINGDOM.

The Highways Agency's business is to manage and maintain the motorway and trunk road network and deliver the Government's programme of trunk road improvement schemes within the policy and resources framework set by Government Ministers. The 10,400 km network for which the Agency is responsible accounts for less than 4% of the total mileage of roads in England. However, the network is the backbone of the country's transport system and is used for about a third of all road journeys and about half of all lorry journeys.

It is also a network that is under pressure. While the Government wishes to reduce the need to travel and influence the rate of traffic growth, demand could still double by 2025. Business users, private motorists, and others who use the strategic road network want a better, safer, and more reliable service. Yet concern about the environmental effect of traffic and of road building is growing. All the Agency's activities have to strike a balance between the environmental consequences and the economic, environmental, and safety benefits of improving traffic flows and removing through traffic from unsuitable roads in towns and villages. Achieving this balance to meet the needs of road users has to take account of the reducing availability of public funds for roads. While the Government's Private Finance Initiative (PFI) cannot meet all future investment needs, it is a vital part of the Agency's strategy to provide the best possible service to road users.

The PFI was launched in November 1992 and while the label was new, the concept - certainly for the Highways Agency - was not. The first major British project to be developed in accordance with the principles of the initiative was the third crossing of the River Thames at Dartford to the east of London. This was a conventional BOT project where a concession was awarded to a private consortium to design, build, finance, and operate a new crossing (the Queen Elizabeth II Bridge) and the existing tunnels, and to recover the cost by levying user-paid tolls. Contracts were signed in 1987 and the new bridge opened in 1991, doubling the capacity provided by the existing tunnels under the river and delivering much needed relief for traffic on the M25 London Orbital Motorway.

The project is financially free-standing. The only public sector contribution was the cost of the approach roads and the transfer of the revenue from the existing tunnels to the concessionaire. The concessionaire competed for the right to design, build and operate the crossing, and to make a return on the investment by charging tolls. The concession period is for a maximum period of 20 years and toll levels are regulated because of the near monopoly nature of the concession. This model was adopted again to provide a second crossing of the River Severn between England and Wales which opened to traffic in June 1996.

With the exception of a small number of tolled river crossings on strategic routes, there is no tradition in the UK of charging users of roads at the point of use through direct tolls. Neither is there legislation in place that would enable user-paid tolls to be introduced on all or part of the existing motorway or trunk road network, although there are powers for new tolled motorways to be built and operated by the private sector. The Birmingham Northern Relief Road is being taken forward as a financially free-standing BOT project in this way. The policy option of bringing forward legislation to authorise the introduction of user-paid tolls on existing roads remains open and trials of electronic tolling equipment are being carried out. One of the previous Government's policy objectives in deciding to proceed with DBFO projects, was to encourage the development of a private sector road operating business in the UK prior to the possible introduction of user-paid tolls across the motorway network.

Prior to the introduction of the PFI, the traditional approach to procuring road improvement schemes on overland routes was to let construction contracts, either built to the Agency's design, or under the Agency's design and build contract. The contractor was paid by the Agency from public funds on achievement of agreed milestones reflecting the work carried out. But, in both cases, the operational and maintenance responsibilities for the capital asset rested with the Agency and the public sector. A further disadvantage of these traditional forms of contracting was that the relationship between the Agency and the contractor tended to be adversarial. Once a contract was let, the contractor would claim for additional costs incurred on the basis that the assumptions on which the price was set were incorrect. For example, claims would be made if ground conditions were not as specified or, if work could not be completed to timetable because of bad weather. This often resulted in time overruns and additional unbudgeted costs for the Agency.

6.1 *The objectives for DBFO projects*

The Agency launched its use of the PFI to procure a road service in August 1994 having previously undertaken a market sounding exercise and consulted likely private sector participants. The Agency's objectives for each DBFO project were to:

1. minimise the contribution required from, and to optimise the extent of the risk borne by, the public sector;

2. ensure that the project roads were designed, maintained and operated safely and satisfactorily so as to minimise any adverse impact on the environment and maximise benefits to road users;

1. promote innovation, not only in technical and operational matters, but also in financial and commercial arrangements; and

2. to meet the Government's objective of fostering the development of a private sector road operating industry in the UK.

The DBFO concept of a road service includes assuming responsibility for the operation and maintenance of lengths of existing road and ensuring that specified new construction works along the length of road are built and made available for road users.

The main benefit of this arrangement is that, by transferring responsibility to the private sector for designing, constructing, financing and operating roads, and increasing payment levels when road construction works are completed, the private sector starts to look at its obligations as a whole, over the 30 year life of the contract, taking full account of the risks inherent at each stage of the project. For example, there is a significant relationship between the way a new road scheme is designed and constructed and its operational costs. The private sector has to make decisions about how it will provide the service to the level specified by the Agency. Allocating those project risks which it is capable of managing to the private sector, results in a lower whole life cost for the Agency, and an improved and more efficient service for the road user.

The DBFO contract fixes the design of the new works elements which the winning bidder has proposed, specifies a date for completion of construction works and the core requirements with which those works must comply, and specifies the operational service requirements for the improved length of road and any existing roads included in the project. In return, the DBFO company receives regular payments from the Agency throughout the life of the contract, based primarily on the number and type of vehicles using the project road.

The eight DBFO contracts awarded to date are for 30 years. That period was selected partly because the construction costs of a project were likely to be financed predominantly by third party debt. Market research amongst potential bidders confirmed this view. Finance for this type of project generally has a maximum repayment period of around 20 years and the payment mechanism had to be structured to allow repayment of debt over a similar timescale.

6.2 *Payment mechanisms and incentives*

The DBFO company receives payments for providing the road service over the lifetime of the contract based primarily on traffic levels and with different "shadow tolls" payable according to the number and type of vehicles using the road. There is a ceiling on the amount of "shadow tolls" payable to ensure that the Agency's maximum liability is capped. However, the DBFO contract is structured so that DBFO operators can also earn performance bonuses or incur deductions from the "shadow toll" payments if the road is not available to users or its use is restricted.

To expand on the payment structure, the concept of a usage/demand payment in PFI contracts is reflected in the "shadow tolls" element of the revenue stream. Different payments are payable for traffic within different traffic bands and dependent on the length of vehicle. Bidders were asked to bid the parameters of traffic levels for a maximum of four bands, with the proviso that the top band must have toll levels set at zero. Capping the Agency's liabilities to DBFO companies in this way assists with internal budgeting and minimises the risks of the private sector making excessive profits at the expense of the taxpayer. Within each traffic band the bidders specified a toll for two categories of vehicle - vehicles over 5.2 metres long (which includes heavy goods vehicles) and those less than 5.2 metres. Bidders set the bands and tolls from their own assessment of traffic levels. Most bidders opted for four bands with the lower band representing existing traffic and with tolls in the lower band set at a level that would cover debt service requirements (but would not provide a return on equity). The result was that the lower band gave funders an adequate payment stream if traffic volumes were not to increase above existing levels.

The PFI concept of an availability payment applies where the project road consists of an existing stretch of road with one or more new construction schemes along its length. In that situation, "shadow toll" payments are made at a reduced level representing the availability of the existing road. This level varies substantially depending on the nature of the DBFO project. In the case of DBFO projects which have no lengths of existing road and the contract is concerned entirely with new construction, no payment is made until the Permit to Use is issued and the road opened to traffic..

Each construction scheme has a shadow toll profile attached to it. Once the Permit to Use is issued for a construction scheme, the DBFO Company receives 80% of that profile. Once the Agency has issued the Completion Certificate, the DBFO Company receives 100% of the scheme's profile.

There are also performance payments and penalties within the DBFO contract. One of the Agency's key objectives is to reduce accident rates on the trunk road network and operators are encouraged to propose safety improvements which, if adopted, enable them to earn bonuses. The DBFO company builds and pays for the safety scheme and is recompensed by receiving 25% of the economic cost of each personal injury avoided in the following five year period. Injuries avoided are determined by comparing the actual statistics with data over the previous three years.

Disturbance and delays caused by lane closures are a major concern to road users and operators incur financial penalties if the use of the road is restricted. A deduction is made from the "shadow toll" payment, the size of which is dependent on the number of lanes closed and the time of the closure. This financial incentive ensures that operators consider the needs of road users and that users are inconvenienced as little as possible by roadworks.

One of the main reasons for using shadow tolls as the primary method for remunerating DBFO companies, was to acclimatise the private sector to the concept of payment per vehicle as a stepping stone to the possible introduction of real toll roads. There are good reasons for paying by usage and it is not inconsistent with the wider aims of the Government to reduce the need to travel, optimise the use of the existing network, and encourage greater use of alternative modes of transport. However, the Agency is exploring moving away from a revenue stream predominantly determined by usage because it wishes to find alternative mechanisms which focus on the performance of the road and incentivises DBFO companies to achieve the Agency's performance objectives while continuing to deliver value for money.

The current lane availability charges and safety bonuses are two areas of operational importance which have payments (or deductions) linked to them. The Agency has developed this concept in its latest DBFO project for the A13 Thames Gateway which will provide the vital transport infrastructure to support the regeneration of East London and Docklands. This is the first urban DBFO project and the first project to reflect the current Government's integrated approach to transport by incentivising the private sector to deliver a service in tune with the five fundamental criteria of integration, accessibility, safety, economy, and environmental impact. This is being achieved through a combination of developing a new payment mechanism and changing some of the core requirements in the contract. The payment mechanism focuses payments on usage by public transport and goods vehicles rather than cars so that the DBFO company concentrates on managing the road for these vehicles and has no incentive to increase car commuting.

6.3 *Service specification*

For the first DBFO projects the scope for the bidders to innovate and achieve cost savings was limited to the detailed design of the new construction elements of the projects, because all the road schemes had completed the statutory approval process. It was a core requirement of these initial contracts that the bidders comply with the undertakings given by the Agency at public inquiries, and that the final design accord with the approved statutory orders. The Agency also provided standards which would be one option for meeting the core requirements. These illustrative requirements were not mandatory and included the Agency's own design manuals and its own proposals for junction layout and design of structures.

6.4 *Expiry of the contract*

Throughout the life of the DBFO contract the Secretary of State for the Environment, Transport and the Regions retains ownership of the road and underlying land with the DBFO company occupying the road and land under a licence. At the end of the term the project road and all fixed facilities on it must be handed back the Agency in a satisfactory condition, ie. the project road must have at least a 10 year life expectancy on handback.

Towards the end of the term there is a mechanism for inspection and agreement about the action needed to ensure that the handback criteria are met. If necessary, the Agency can withhold "shadow toll" payments to the DBFO company to ensure that if the road is handed back in a substandard condition there is no cost to the taxpayer of rectifying faults.

6.5 *The Public Sector Comparator*

To ensure that bids satisfy the twin tests of delivering value for money and optimal risk allocation, the Agency prepares a public sector comparator before inviting bids for the projects. The public sector comparator is calculated by costing what the public sector would have had to pay to procure the same underlying asset by traditional means, and the cost to the Agency of operating and maintaining the road over 30 years. A value is also added to represent the cost of the risks adopted by the private sector and priced for in their bids. Unless there are special circumstances, the DBFO contract is only let if the NPV of the final agreement between the Agency and the preferred bidder is lower than the risk adjusted public sector comparator.

The public sector comparator has shown that, by using DBFO arrangements to deliver new road schemes and the on-going operation and maintenance of the motorway and trunk road network, the Highways Agency is obtaining value for money savings averaging 15 per cent across the first eight projects.

6.6 *Risk allocation*

Project risks should only be transferred to the private sector if, and to the extent that, the private sector is capable of bearing such risk. DBFO contracts transfer to the private sector a substantial degree of responsibility for constructing, operating and maintaining the project road and for financing the relevant costs. Transfer of responsibility increases the scope for innovation by the private sector. The risks associated with those obligations are, in most cases, transferred to the private sector so that if the specified service is not provided the Agency pays no more than agreed at the outset.

The Agency carried out an analysis of the risks attaching to each project by drawing up a risk register which set out in detail the risks relevant to each stage of the project, the likelihood of those risks occurring, and an estimate of the financial impact of occurrence. This analysis helped the Agency to establish the type and the quantum of risk they should ask the private sector to take. The DBFO contract is drafted so that the DBFO company bears all risks associated with an area of delivery, such as operation, unless the Agency is specified to take a risk, either through the payment mechanism, change mechanism or termination events. Therefore, any unanticipated risk is borne by the private sector. This underlines the importance, for both the public and private sectors, of undertaking a detailed risk analysis for each project.

The concept of the PFI is that, the private sector will generally be asked to take the risk of increased construction and operation cost overruns, the risk of delay to delivery of the service, the risk that the design of the underlying asset will not deliver the required service, and the change of law risk (other than any change of law which specifically discriminates against PFI companies). DBFO contracts are structured in this way. The allocation of risks which are unique to DBFO roads include, traffic risk, protester risk, and latent defect risk. The handling of these in DBFO contracts is covered in greater detail below.

6.7 *Traffic Risk*

This is a key area of risk transfer in DBFO projects because traffic risk is very closely allied to revenue risk. The number of vehicles using the road affects the cost of constructing it with a reasonable life expectancy and of maintaining it to the required standard. The Agency has its own updated traffic projections for each project road which are based on the initial investment decision for road schemes within the DBFO project. These forecasts are kept confidential because of the objective of fostering the development of a private sector road operating industry with the skill base necessary to take strategic decisions on road design, construction, operation and maintenance. A key skill is the ability to forecast road usage in the future. The Agency also uses it's confidential forecasts during the evaluation of DBFO bids to calculate how much risk each bidder's payment structure would place on its revenue stream at the Agency's "most likely" traffic forecast.

If bidders' traffic forecasts under-estimate traffic they will weight most of their return in the lower bands and therefore, their payment stream will have little risk attached resulting in poor value for money. If, however, the bidder overestimates traffic growth and the return is weighted in the higher bands, their payment stream will have more risk and will provide better value for money. But, there may be a point at which a bidder's assumptions on traffic growth are so optimistic that, in the Agency's view, the bidder's proposed structure is financially unstable.

6.8 Protestor Risk

In the UK, a growing awareness of environmental issues has resulted in a vociferous anti-roads lobby, sections of which resort increasingly to direct action in attempts to frustrate and delay the construction of new or improved roads. This has given rise to significant extra costs for the Agency. There is a possibility of disruptive protestor action on some DBFO projects and the Agency recognised that value for money could be damaged if the bidders were asked to bear all the costs. However, DBFO companies would be in control of the works and therefore, in the best position to determine the strategy to deal with problems. For this reason, bidders were asked to price options ranging from accepting 100% of the risk to none of it. The end result of negotiations was generally a project specific, risk sharing arrangement between the Agency and DBFO company providing encouraging evidence of bidders' willingness to accept new types of risk.

6.9 Latent Defect Risk

Where a DBFO project involves the DBFO Company taking over responsibility for operating an existing length of road, bidders have had to take a view on the state of the road and any structures on the road. Their technical advisers have carried out investigations but there may be problems which cannot be detected or "latent defects" such as spalling of concrete or a structure component not meeting the expected design life. As with protestor risk, bidders were asked to bid on options ranging from 100% acceptance of risk to zero. The outcome of negotiations in all cases was that the preferred bidder accepted 100% of the

latent defect risk. In future DBFO contracts, the allocation of risk for latent defects will not be a biddable option - bidders will be required to bid on the basis that they accept 100% of the risk unless there are specific and exceptional circumstances which indicate that better value for money might be obtained through some sharing of the risks.

7. CONCLUSIONS

The Highways Agency has an impressive record of delivering innovative PFI projects through its DBFO programme. For many major projects, the traditional method of contracting with private sector contractors building a length of road to the Agency's design and then passing responsibility for operating and maintaining it back to the public sector has been overtaken by DBFO arrangements.

By transferring construction and operational risks to the private sector and by harnessing the private sector's entrepreneurial skills, the Agency has also achieved better value for money for the taxpayer than if the projects had been procured conventionally. Because DBFOs can deliver significant value for money benefits, they are now the procurement method of choice for many major road improvement projects. The Agency is no longer buying a capital asset but buying the provision of a road service for 30 years on behalf of the road user.

Contracts have been awarded for eight DBFO projects. These involve the delivery of new road improvement schemes with a capital value of over £500 million enabling much needed road schemes to start earlier than if they had been dependent on conventional public sector funding arrangements. They also provide for the operation and maintenance of over 600 km of existing road. The first urban DBFO project involving about £146 million of road construction work is also being developed and invitations to tender for this project are scheldued to be invited during 1998.

Making the most efficient use of the motorway and trunk road network, and providing the best possible service to users of that network, is an immense challenge. The DBFO programme and other public/private partnerships will not meet all the future investment needs. But, pursuing private finance opportunities and harnessing the skills and expertise of the private sector is likely to remain a vital part of the Highways Agency's strategy to deliver the best possible service to users of England's strategic road network.

Operation and Maintenance of Large Infrastructure Projects, Vincentsen & Jensen (eds)
© 1998 Taylor & Francis, ISBN 90 5410 963 7

World-wide study of best practice within railways: Technology Whitebook

P.I. Koch
ScanRail Consult, Copenhagen, Denmark

ABSTRACT: This paper presents the major themes derived from the Technology Whitebook Study performed by ScanRail Consult for the European Commission during the period from February 1997 to November 1997. The Technology Whitebook Study is a management and market analysis of the European Railways presenting some visionary strategies for railways to enhance their competitiveness touching upon managerial, organisational as well as technical aspects.

1. A VISION FOR THE EUROPEAN RAILWAYS

In the year 2010, European railways are healthy and profitable businesses operating in a free competitive market driven by customer and market demands. They are international business-orientated companies with focus on the customers. The European railway operators are trans-national and intermodal transport providers offering seamless transport services covering all four transport modes - water, air, road and rail. Railway companies are now fully deregulated and privatised.

Market share for passenger transport have increased from approximately 7% to 15% over the last 10 years. During the same period, the market share for freight transport has increased from approximately 15% to 25%. The aim is to gain another 10% of the market for both segments within the next ten years. To achieve this substantial initiatives have to be taken:

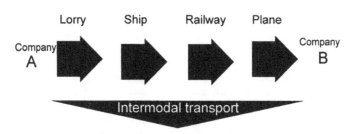

The total transport solution

Figure 1. The total transport solution 1

Figure 2. The substantial initiatives.

The overall vision for the new European companies can be summarised as:

Operators
are market leading trans-national transport companies offering world class intermodal/ seamless transport solutions within freight business and passenger transport - business as well as leisure - throughout Europe.

Infrastructure Managers
are the market leading Facility Managers comparable to airports. The infrastructure manager is not just a time slot provider but is acting as a facility manager conscious about attracting operators and customers to the railway. The traffic is controlled on a trans-national basis.

The railway industry, which plays a central and increasing role in the total European transport system, is characterised by:

- Modern organisations accommodated to act business oriented, with committed staff, customer focus and ability to gain market shares and be profitable.
- A formalised network structure consisting of a number of trans-european lines with a certain level of standardisation to ensure interoperability in Europe. The network is respectively dedicated for freight and intercity passenger traffic.
- International operators offering seamless transport-services Europe-wide. Several operators work in intermodal alliances with other transport providers to ensure door-to-door services.
- A pan-European centre for awarding and planning time-slots for international transport on main corridors.
- The railway players have increased their buyer power towards suppliers of rolling stock and infrastructure equipment by implementing EU standards based on well-defined functional requirements and open interfaces.

1.1 Framework for the vision of the European railway industry

Since the late nineties the development of the railways have changed focus from technical aspects to more business driven aspects, and both the railway companies as well as the European Commission now places high emphasis on market orientation and commercialisation.

The Commission has been very conscious about creating free access to the infrastructure and have succeed in getting all the countries in the EU to comply with the EU-directive 91/440. This means that the national railway companies are now split up in an infrastructure manager company and a number of railway operators. Eventually, the following key areas constitute the framework for the vision for the European railway industry:

1.2 Profitability, Financing and Debt

European Rail companies are in 2010 profitable and healthy corporations listed on the stock exchange. Therefore it is no problem to attract private investors for investment in new rolling stock, for construction of new lines and upgrading of existing lines as long as these investments are based on commercial calculations and provides a reasonable pay back. The return on investments are approximately 10%.

1.3 Products, services and Corporate Image

Customers perceive the level of competence of the new rail companies as best practise in every respect - both in terms of customer orientation, products, corporate image, precision, quality, information, flexibility and price. Rail companies are perceived constantly innovative in terms of developing new customer oriented door-to-door intermodal transport services. This is particularly crucial for export oriented companies heavily dependant on a competitive transport sector. The corporate image of the railways has been substantially improved and is now comparable to that of aviation.

Focus on business processes
Since 1997 initiated by the EU commission Rail companies have continuously used Business Process Reengineering supplemented by other commercial/business improvement methods for improvement of work and business processes. Always taking starting point in customer and market demands.

1.4 Personnel and Management

Railway companies now subject to the same agreements of employment as the competing modes and have the same opportunities to hire and dismiss labour as have air and truck companies. Thus, railway companies are setting forth demands on unions and are able to influence decision making processes which has lead to agreements with unions on removal of rigid skill demarcations. Eventually, railway companies are now employing a flexible and multi-skilled labour force consisting of both old railway people and new people with commercial backgrounds.

1.5 Shift of role in the value chain

The new rail (intermodal transport) companies have integrated forward in the supply chain and are now well estimated business partners. Rail companies act as integrated transport/logistic partners - as invaluable facilitators of the customers value chain - offering reliable and best practise solutions to multinational companies as well as individuals. Customers with specialised transport need or large shipments have in the same way integrated (backwards) the transport service into their own business where it is profitable and by that become rail operators themselves.

1.6 Liberalisation/privatisation

The positive development of the railways have attracted third party investors, resulting in the operators being fully privatised, while the governments maintain a majority of the shares of the infrastructure manager, ensuring public control. The liberalisation process has in a positive way separated railways from political interference, and has thereby increased the possibility to make autonomous decisions, which may include less risk averse strategies.

1.7 Supplier industry

The railway players are competing and co-operating simultaneously to strengthen common issues of interests that make them collectively more efficient and strong as an industry. This has lead to an increased buyer power towards suppliers of rolling stock and infrastructure equipment by implementing common standards, which in turn increases the internal competition between railway suppliers and lowers the costs.

1.8 New structure of industry

All companies act internationally and are strengthened through mergers of smaller national companies. Rail companies have in the late nineties faced a dramatic challenge in terms of competitors from other modes as well as from new European operators and US operators entering Europe. The process involves the whole industry including the European Commis-

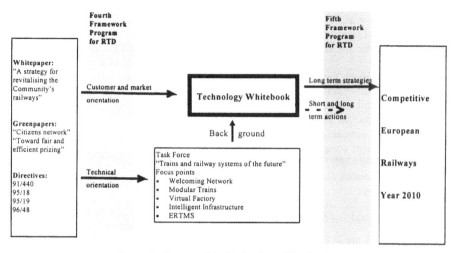

Figure 3. Context of the Technology Whitebook

sion, national ministries of transport, suppliers, operators and infrastructure managers.

2. APPROACH

2.1 Preconditions and framework

The Technology Whitebook has evolved within the context of already existing directives, documents, RTD-programmes and actual implementations of improvements efforts, all made to enhance the competitiveness of railways. The starting point of the Technology Book has been a customer point of view. The purpose of the Technology Whitebook is to link the needs of customers to the strategies for technical, organisational or economical improvements in order to create competitive railways accomplished to meet and react on market demands. The context of the Technology Whitebook is shown in figure 3.

2.2 The basic approach

In Figure 4 the basic underlying approach for the analysis is illustrated. The analysis is based on a market approach identifying major threats and opportunities and the influence of customers and competitors on the competitive situation of railways. The identification of the competitive situation leads to formulation of strategies for reengineering European railways -translating them into competitive railways in the year 2010.

3. EXECUTIVE SUMMERY

In the last four decades the railways have experienced a continuous loss in market shares. The White Paper "A Strategy for Revitalising the Community's Railways", presented by the European Commission in July 1996, emphasises that there is a need for an increased competitiveness of the European railways if they are to survive within the next 2-3 decades. The European Commission has wished to follow up the work on the White Paper in order to support the strategies proposed in the paper. This was the main reason why the Technology Whitebook, containing relevant technological, organisational and international strategic issues and action plans, was invited for tender.

In other words, there is an urgent need for fundamentally restructuring European railways and the way they function today, changing the railways from a supply orientated situation into one of market structure and enhanced efficiency. However, there are railways around the world which not only maintain their market shares but which also operate profitably. The purpose of this Whitebook is to analyse the main issues and problems facing European railways and to gather inspiration from the success stories outside Europe, in order to propose strategies for drastically increasing the competitiveness of the railways.

The overall *vision* for the European railway can be summarised as: The operators are the market leading transnational transport companies offering world class intermodal/seamless transport solutions.

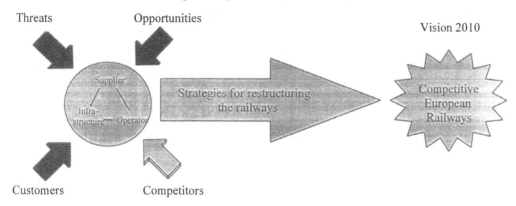

Re-engineering the European Railways

Threats Opportunities

Vision 2010

Supplier

Infra-
structure — Operator

Strategies for restructuring
the railways

Competitive
European
Railways

Customers Competitors

Figure 4. The basic approach of the Whitebook

And the infrastructure managers are the market leading Facility Managers comparable to airports and conscious about attracting operators and customers to the railway.

The first part of the report, containing an analysis of the present status of the competitiveness of the European railways, points at three important measures to identify the competitive situation of the industry - namely market share, market position and profitability.

It is found that the *market shares* for railways have suffered a decrease in spite of the fact that the need for transport has experienced a constant increase in the last 30-40 years. The analysis points to the fact that the major part of the increase in the transport market has gone to road transport. The major strategic objective is thus found to be one of selectivity of market segments and focus strategies.

The analysis of *market position* strongly indicates that the main competitive parameters to embark on for the railways to improve their market position are price, time, quality, flexibility, door-to-door services and in addition comfort for passenger service.

The third issue in the status of the competitiveness section - *profitability* - shows that it is crucial to implement a bottom-line approach for railways to become profit oriented and to improve efficiency. Also, cost consciousness and transparency in costs are important to increase profitability. Thus, a break with traditional structures of deficit covering and subsidies is an imperative. An important tool is found to be the introduction of public service contracts - making society a customer - which is a major incentive for railways to shift from supply oriented towards market oriented organisations.

By looking into railways in other parts of the world - US, Japan, New Zealand and Europe - it has been possible to capture *best practise* experiences and synthesise them into important lessons to learn.

In the *US* the railways are highly dominated by freight business. Some of the key areas are a general tendency in strategies to embark on longer distances, longer hauls, and higher axle loads within freight. Moreover, tight cost control, conscious pricing strategies, introduction of business process reengineering methods, strong business orientation and aggressive marketing characterise the industry. In terms of RTD programmes these were to a high degree commercially oriented.

The study of the *Japanese railway* shows that the national railways have been successfully privatised which have led to improvements in operations and productivity and have changed the railway into a profitable business. The dedicated infrastructure for high speed trains ensures a high punctuality, reliability and utilisation of capacity. A continuous effort to reduce maintenance costs has led to the development of a new almost maintenance free commuter train with a lifetime of 13 years. A strong focus on affiliated business contribute significantly to revenues and represents 20% of total revenues.

The study of the railway in *New Zealand* likewise pointed to the positive gains from commercialising and privatising railways. Also, the study indicated the importance of pursuing an aggressive marketing strategy by means of market segmentation and customer analysis. The identification of core and non core businesses has proven essential to increase profitability. Eventually, the perception of railway transport as an integrated part of a broader transport chain emphasising seamless transport and the door-to-door concept characterises the railway.

One of the main experiences from *Europe* is that there is a lack of standardisation in the supplier industry leading to unnecessary high costs of rolling stock and equipment. A shift from detailed towards common functional requirement specifications will result in increased standardisation and lower prices. The main competitor to the railways is the truck industry from which market shares should be gained. A milieu of competition and co-operation between railway operators - also at an international level, e.g. in terms of strategic alliances - is desirable. Furthermore, the idea of a dedicated network for freight and passenger transport should be further developed. Eventually, RTD projects should be commercially oriented and applicable to the railway industry to a much larger extent - and not only carried through due to pure scientific objectives

3.1 Case example - the Approach to an Integrated Maintenance system

The Japanese have developed a new Computerised Safety, Maintenance and Operation System called COSMOS. COSMOS consists of several functions that incorporate the management of operation and maintenance. The functions are:

- traffic management system including train schedule planning
- traction power control system
- maintenance work management system

A unified systematic flow of the work processes has been obtained, which has resulted in a time reduction of the work processes. For example the train schedule planning system includes a function for directly setting up train schedules, crew and rolling stock assignment, rolling stock inspection and operating schedules. This was previous made manually and it is made more efficient with the introduction of COSMOS.

The inspection of the infrastructure is carried out by the 'Dr. Yellow train set', which is a multipurpose inspection train that plies the line at 210 km/h, inspects the condition of tracks and electrical facilities every 10 days. Onboard instruments collect and computer analyse data, which is used in the planning of maintenance work and also gives an early detection of potential sources of trouble. When developing new Shinkansen trains the focus was on high speed. In the future the reduction of the maintenance cost of rolling stock will play a major role.

3.2 Punctuality and capacity utilisation

An important factor for the high punctuality and capacity is the integrated use of computer systems. Development of Computer-Aided Train Control (COMTRAC) and the new Computerised Safety, Maintenance and Operation system of Shinkansen (COSMOS) plays a major role in the success.

3.3 Approach to RTD

The focus of research and development is on:

- saving maintenance
- saving energy
- environmental improvement (noise reduction)
- improvement of rush hour congestion

4. OVERALL FINDINGS

Some of the most important overall findings from the international studies of railways can be summarised as a tendency going in the direction of corporate transformation and business development which is *commercialisation* with a strong focus on market segmenting and identification of comparative and competitive advantages easing the way to privatisation of railways; the establishment of a *dedicated network structure* for *freight* and *passenger* traffic; and increased intermodality allowing for *seamless transport* and integration between different transport modes like rail, road, air and sea.

The analysis of *other industries* looking into the air, sea, energy and telecommunication industry points at strong parallels to the development going on in the railway industry in terms of processes of liberalisation, privatisation and mergers and acquisitions. It became evident that successful development can be

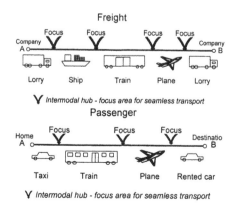

Figure 5. Intermodal Transport

Figure 6. Example for focus areas in rail transport development - European Freight Freeways

achieved by introducing the limited liability / public liability company model in earlier state owned companies. Important lessons are the pursuit of strategies to meet free competition, in terms of e.g. business process reengineering to streamline organisations, and strategic alliances to increase internationalisation and to capture economies of scale.

In the second part - the synthesis - The proposed strategies are separated into *General strategies* which contain overriding issues that must be handled on a European level and *Corporate strategies* which contain specific issues that concern the individual operators and infrastructure managers.

The general strategies consist of three main issues. The first one is establishing *Pan-European railway networks* consisting of main corridors dedicated for passenger or freight transport respectively and with high level of interoperability. To optimise co-ordination and strategic planning of international traffic a European Time Slot Administration Centre is introduced.

The second issue concerns *corporate transformation* of existing railway companies and deals with

business orientation of the infrastructure manager, establishment of a competitive environment for European operators, preconditions and a legal framework for corporate transformation.

The third issue concerns ways of *reducing cost* in the railway industry and describes the establishment of a competitive environment for suppliers of standardised equipment. Also, improved utilisation of assets and methods for financing and easing new operations are imperatives.

In corporate strategies the objective is to present strategies to improve business for operators in various market segments and for infrastructure managers. The strategies are fully company-oriented and can be implemented independently in the different railway companies.

Finally, the analyses and strategies forms the background for Recommendations for focus areas in rail transport development projects the next 1 - 6 years.

CONCLUSIONS

The work presented in this paper indicates that focusing exclusively on technological aspects (best practise, information technology) is not sufficient to save the railways. It is thus of increasing importance that the railways use Management, their strengths and a mix, in order to co-operate and provide stronger and more cost efficient passenger and freight services, by offering seamless national as well as international transport.

The structure and conclusions of the present paper are illustrated in the following figure. Additionally a picture of the path of how to realise the vision, is sketched out. The path is based on the issues outlined in the strategies.

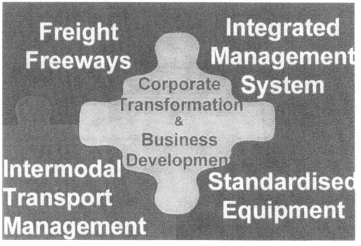

Figure 7. Examples of Corporate transformation and business development

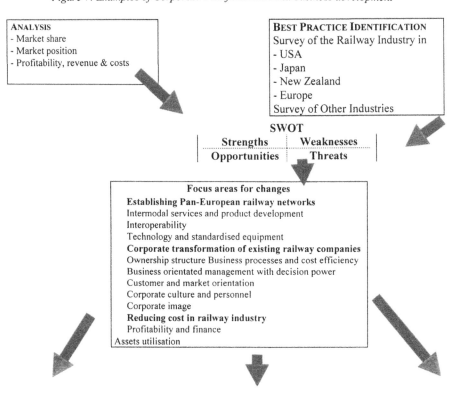

COMPETITIVE SITUATION

Common European issues

Vision

In 2010 the railways are characterised by:

Business orientation supported by a common European infrastructure network for freight and long distance passenger transport and a well functioning commuter traffic.

The European network is connecting major industrial and societal junctions, allowing operators to draw full advantages of the railways competitive strengths and react to the needs of the customers for an entire transport package.

The dependence of specific suppliers has been minimised.

Inter City Traffic

Strengths
- Short travel time from city centre to city centre
- Use of travel time
- Cheaper than air transport

Weaknesses
- Long travel time on distance more than 1000 km
- Lack of focused customer products
- High investment in rolling stock
- Fixed pricing

Urban / Suburban

Strengths
- Lower costs in peak hours
- Less affected by rush hour
- High political and public support

Weaknesses
- Lack of door-to-door service
- Poor utilisation outside rush hour
- High capital and operating costs

Freight

Strengths
- Low costs for large scale
- Decreasing marginal costs
- Low personnel per tonnage ratio

Weaknesses
- Low flexibility
- Slow to react on needs of customers
- Poor service level
- High fixed and start-up costs
- Poor utilisation of wagon capacity

Infrastructure

Main issues
- No incentive to improve performance
- Poor customer focus
- Low return on investment
- Strong political interference

CORPORATE STRATEGIES

Inter City Traffic
- Focus on traffic from metropolitan to metropolitan.
- Focus products to customer segments
- Aggressive marketing
- Profit-maximisation approach
- Strategic intermodal alliances
- Night traffic

Urban / Suburban
- Operating on social contracts
- Improved intermodal connections and customer orientation
- Better utilisation of assets
- Reduce personnel cost
- Profit-maximisation approach

Freight
- Focus on long distance and large volume approach
- Focus on short networks near to customers to act as feeder networks
- Change focus from product to customer
- Full service provider
- Surveillance of freight
- Minimise empty returns
- Second hand market of rolling stocks and the role of renting

Infrastructure
- The infrastructure manager as a facilitator
- More focus on affiliated business
- Utilisation of the infrastructure capacity
- More structural and automatic surveillance and planning of maintenance
- Life cycle cost reduction
- Differentiated quality of infrastructure used for passenger and freight traffic

---

General Strategies

Establishing Pan-European railway networks
1. European network consisting of main corridors.
2. Establishing dedicated networks for passenger and freight.
3. European Time Slot Administration Centre.

Corporate transformation of existing railway companies
4. Business orienting the infrastructure manager.
5. Establish a competitive environment for European operators.
6. Establish a legal framework for corporate transformation
7. Establish the preconditions for corporate transformation and privatisation

Reducing cost in railway industry
8. Establish a competitive environment for suppliers of standardised equipment.
9. Improved utilisation of assets
10. Methods for financing and easing new operations.
11. Cost efficiency measures

Path towards realising the vision

2010

The vision described previously has been fully realised.

2005

The technical issues of the European network for passenger and freight transport have been solved and agreed upon. A large part of the network is operational for inter-European use, while there is an ongoing process of upgrading the main corridors which are still not up to date.

2002

EU-standards regarding functional specifications and open interfaces on supplier equipment have been formulated. The level of interoperability on the main corridors has been decided and has been implemented where technically possible. The infrastructure has been opened up for third party financing, mainly because of transparency of the cost structure and clarification of the services and products offered.

1999

International co-operation between the European countries has been established with regards to operating trains across European borders. There is an ongoing research programme focused on corporate transformation and solving the technical compatibility issues necessary for total interoperability. Some of the main corridors have been dedicated for either freight or passenger transport. The interface between infrastructure managers and operators is clearer and the cost structure of the infrastructure is more transparent.

1997

EU, major confederations within the railway sector, governments and national Ministries of Transport have initiated major initiatives to drastically improve the poor competitive situation of European railways. These initiatives includes short term actions and substantial research programmes in order to realise the vision of an efficient and profitable European railway. Railway companies have started to see themselves as transport providers and the corporate transformation processes concerning business orientation is taking speed.

Operation and Maintenance of Large Infrastructure Projects, Vincentsen & Jensen (eds)
© 1998 Taylor & Francis, ISBN 90 5410 963 7

Organization model for operation and maintenance of major state owned toll bridges in California

Thomas M. Rut

California Department of Transportation (Caltrans), Office of Structure Maintenance and Investigations, Sacramento, Calif., USA

ABSTRACT: Toll bridges in California are generally landmark structures in highly congested urban areas. They are very visible structures with high traffic volumes that generate significant public revenue. Operational problems generate substantial media attention. This media attention, the large revenue, and high visibility makes these structures political targets and the resulting legislation affects the operation, maintenance and rehabilitation of these bridges. The 9 state toll bridges average 38 years of age and require constant maintenance and rehabilitation. They are currently undergoing extensive earthquake retrofitting. The organizational structure and responsibility breakdown for operation, maintenance, and rehabilitation are very complex and affected by all of the above factors. This paper will delineate the organization that is in place, discuss some of the issues faced, and forecast some future concerns and expectations.

1. INTRODUCTION

Toll bridges in California are generally landmark structures in highly congested urban areas. They are very visible with high traffic volumes that generate significant public revenue. Operational problems generate substantial media attention. In October 1989, when one span of the San Francisco Oakland Bay Bridge collapsed during the Loma Prieta Earthquake, its picture made the national news for weeks on end. Daily radio and television traffic reports detail normal rush hour conditions. Traffic congestion caused by maintenance or construction activities is widely reported in all local media. Recently, a prominent state legislator was caught in a traffic jam caused by one of our construction contractors working near the approach to the Bay Bridge. The activity was continuing after the State specified time for removal of a lane closure due to unanticipated problems with the construction. This incident prompted that legislator to publicly call for a ban on ALL daytime construction and maintenance activities that affect public traffic and for the establishment of an area wide traffic manager that would have dictatorial powers to stop or modify any maintenance or construction activity that is impacting public traffic.

Most of the toll bridges are in our severe coastal environment. This salt laden corrosive atmosphere is very hard on structural steel. A continuous painting and maintenance program has kept these structures in relatively good condition, but, with an average age of 38 years, the maintenance requirements are increasing. One 70 year old bridge has outlived its useful life and is in need of full replacement.

Figure 1. San Francisco-Oakland Bay Bridge
Total Length 13.5 km
Main Span
Suspension 704 m
Cantilever truss 427 m
Navigation Clearance
Vertical 67 m
Horizontal 673.6 m
Year Constructed 1936
Cost $77 million

Figure 2. San Mateo-Hayward Bridge
Total Length 10.8 km
Main Span 228.6 m
Twin Steel Box Girder 2941 m
Conc. Trestle 7863 m
Navigation Clearance
 Vertical   41.1 m
 Horizontal 152.4 m
Year Constructed 1967
Cost $70 million

Figure 3. Dumbarton Bridge
Total Length 2.62 km
Main Span 103.6 m
Maximum Navigation Clearance
 Vertical   25.9 m
 Horizontal   61 m
Year Constructed 1982
Cost $70 million

With the continual advancement of bridge standards and increase in traffic demand, coupled with normal deterioration, this bridge inventory always requires rehabilitation and improvement.

2. BRIDGE DESCRIPTION

This paper is limited to the organization model for the operation and maintenance of the 9 State owned toll bridges. The most recognizable bridge in California, the Golden Gate Bridge, is not included because it is owned, operated and maintained by a separate local bridge district. A description of the 9 State bridges follows.

This double deck bridge connecting Oakland to San Francisco is really several connected bridges. It includes suspension spans, a tunnel, a cantilever steel truss, many shorter steel trusses and various approach spans. It provides 5 traffic lanes in each direction.

This bay crossing connects Hayward and San Mateo on the San Francisco peninsula. The twin steel box girder portion is six lanes wide and the concrete trestle four, to be widened to six in the near future. The main span unit of 114.3-228.6-114.3 m is a twin orthotropic box girder of high strength steel (689,000 MPa).

Route 84 between San Mateo and Alameda Counties is the southern most crossing of San Francisco Bay. The approaches have cast-in place concrete slab spans and precast, pre-stressed trapezoidal shaped girders up to 45.7 m in length. The main spans are trapezoidal shaped steel boxes with a lightweight aggregate concrete deck. The center span unit is 70.1-103.6-70.1 m.

Figure 4. Carquinez Bridges
Total Length of Double Cantilever
Truss Unit 1021 m
Main Span 335.3 m
Navigation Clearance
 Vertical   39.9 m
 Horizontal   304.2 m
Year Constructed 1927 West
                      1958 East
Cost $8 million West
       $ 32 million East

These parallel structures carry Interstate 80 over the Sacramento-San Joaquin River connecting the San Francisco Bay Area to the Sacramento Valley. The main double, cantilever truss units provide two navigation spans with a 152.4-335.3-45.7-335.3-152.4 m arrangement. The West bridge provides 3 lanes westbound and the East bridge provides 4

Figure 5. Benicia-Martinez Bridge
Total Length 1.89 km
Main Span 160.9 m
Navigation Clearance
Vertical   42 m
Horizontal   134 m
Year Constructed 1962 and 1990
Cost $25 million
Traffic 6 Lanes

Figure 6. Antioch Bridge
Total Length 2.87 km
Main Span 140.2 m
Navigation Clearance
 Vertical   41.1 m
 Horizontal   121. 9 m
Year Constructed 1978
Cost $34 million
Traffic 2 Lanes

lanes eastbound. There are various ramp and main road approaches of composite plate girders. The West bridge is a riveted through truss, and the East bridge a welded high strength through truss.
Spanning the Sacramento-San Joaquin River between the cities of Benicia and Martinez, it is upstream from the Carquinez Bridge. Originally

constructed for four lanes, widened to six lanes in 1990, a parallel bridge is under design. There are 10 deck truss spans and nine plate girder approach spans.
Spans the San Joaquin River at Antioch, and connects it to Sacramento County. It has a lightweight concrete deck on continuous composite welded corten steel girders.

Figure 7. Richmond-San Rafael Bridge
Total Length 6.5 km
Main Spans 2 @ 326.1 m
Navigation Clearance
 Vertical   56.4 & 41.1 m
 Horizontal   304.8 & 283.5 m
Year Constructed 1956
Cost $66 million
Traffic 6 Lanes

This bay crossing connects Richmond with San Rafael. The main spans are double, deck cantilever trusses of 326 m each. Each deck provides for 3 lanes of one way traffic, (of which 2 are in use).

This bridge connecting San Diego with Coronado Island has three main spans of 201.2-201.2-167.m and is an orthotropic steel plate box girder. There are 26 steel plate girder approach spans and lightweight concrete deck. Shorter approach spans are precast prestressed lightweight concrete I girders with lightweight concrete deck.

This bridge connects San Pedro with Terminal Island in the Port of Los Angeles. The main suspension span are 154.4-457.2-154.4 m with a stiffening truss and a lightweight concrete deck. The steel tower height is 111.25 m above concrete bases.

Figure 8. San Diego Coronado Bay Bridge
Total Length 3.38 km
Main Span 201.2 m
Navigation Clearances
  Vertical   53 m
  Horizontal   152.4 m
Year Constructed 1969
Cost $70 million
Traffic 5 Lanes (one reversible)

Figure 9. Vincent Thomas Bridge
Total Length 1.85 km
Main Span 457.2 m
Navigation Clearances
Vertical   56.4 m
Horizontal 152.4 m
Year Constructed 1963
Cost $26 million
Traffic 4 lanes

3. DEFINITIONS

Critical to understanding this paper is the definition of the following items as used by the State of California:

Operations:
Day to day activities required to provide the service to the user that the facility was intended to provide. Includes toll collection, traffic handling and utility charges.

Maintenance:
Preservation and repair to keep a facility in a safe and usable condition consistent with its designed configuration. This activity offsets the effects of traffic wear, weather, deterioration, organic growth, damage and vandalism. Special or emergency response and repairs due to accidents, storms, earthquakes or other unusual causes are included.

Rehabilitation:
Restoration of a facility that has become inadequate due to deterioration or other causes. Usually applies to structures classified as structurally deficient or functionally obsolete. Rehabilitation can include improvements to the facility to meet new conditions or criteria. The goal of rehabilitation is to provide a facility that can be economically maintained. The difference between rehabilitation and maintenance is that maintenance is simply to keep structures in a safe and usable condition.

Seismic Retrofit:
Rehabilitation aimed at improving the structure to meet the latest earthquake response criteria. We have separated this from other rehabilitation because of the specific legislation that requires it, as well as the immense scope and cost of the projects.

4. TRAFFIC VOLUME

The nine toll bridges served about 280 million vehicle crossings in the 1996/97 fiscal year. That averages 770,000 vehicles per day. Tolls are only collected in one direction. The annual tolled traffic per bridge follows. The actual traffic volume would be about twice that shown as tolled:

	Tolled Vehicle Crossings 1997/98 Fiscal Year
San Francisco-Oakland Bay Bridge	47,356,679
San Mateo-Hayward Bridge	14,815,603
Dumbarton Bridge	11,192,859
Carquinez Bridges	19,064,849
Benicia-Martinez Bridge	17,189,405
Antioch Bridge	1,780,919
Richmond-San Rafael Bridge	10,819,910
San Diego-Coronado Bay Bridge	11,904,583
Vincent Thomas Bridge	5,521,004

FISCAL STATUS
in $1000

	REVENUE*	OPERATIONS*	EXCESS* REVENUE	MAINT* SHA**	SEISMIC RETROFIT
San Francisco Oakland Bay Bridge	52,203	13,072	39,131	4,024	1,333,000
San Mateo Hayward	15,287	3,796	11,491	1,126	136,000
Dumbarton	9,697	2,567	7,130	486	0
Carquinez	23,843	5,012	18,831	455	87,000 ***
Benicia-Martinez	19,162	3,958	15,204	201	106,000
Antioch	2,438	1,321	1,117	42	0
Richmond San Rafael	12,193	2,844	9,349	2,118	335,000
San Diego Coronado	5,921	3,575	2,346	-------	95,000
Vincent-Thomas	3,325	1,812	1,513	857	38,000
	$144,069	$37,957	$106,112	$9,309	$2,635,000

Figure 10.

*   1996/97 Fiscal Year Data
**   State Highway Account
*** 1927 structure to be replaced by other funds

These bridges were mainly constructed and improved with general revenue bonds, backed by the State of California. Repayment of the bonds and interest comes from the tolls. Most of the bonds for this construction have long been retired, so interest and bond retirement costs are relatively low.

The previous chart shows annual revenue to be over $140 million. The extra revenue available after the payment of operation costs is in excess of $100 million per year. All bond costs and bridge rehabilitation and capitol improvement should come out of this fund, but any uncommitted fund of this magnitude becomes controversial with many competing users eyeing it for a bailout.

The routine maintenance costs ($9 million a year) are widely assumed to come from toll revenue but, except for the San Diego-Coronado Bridge, State law requires that the State Highway Account pay for this work. The State Highway Account's main revenue source is the state gas tax. Legislation enacted in 1992 transferred all San Diego-Coronado Bridge revenue to SANDAG, the San Diego Area Regional Transportation Agency, and made this local agency responsible for Caltrans' costs to operate and maintain this bridge. The local agency is also responsible for the cost of all improvements to this bridge and its approaches.

5. SEISMIC RETROFIT

After the 1989 Loma Prieta earthquake, the legislature mandated that all public bridges in the State that didn't meet current seismic standards be retrofitted to improve their seismic capability. This legislative mandate did not address financing at that time. The complexity and large size of the toll bridges have made them the most difficult and costliest component ($ 2.6 billion) of our statewide retrofit program. They will also be among the last retrofits completed. Logic would point to the

ORGANIZATION CHART 1
ENGINEERING SERVICE CENTER

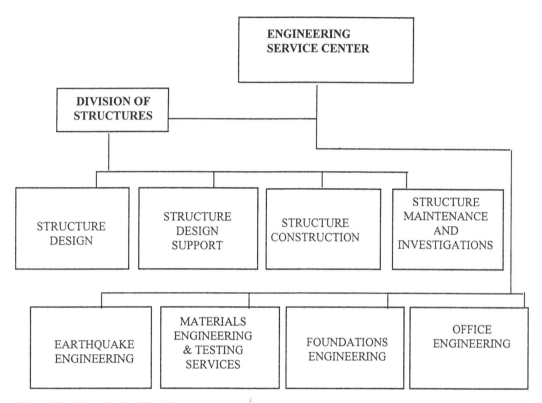

Figure 11. Organization Chart 1

excess toll revenue as a source of seismic retrofit funding for the toll bridges, but previous political promises, voter action and regional competition make a funding compromise almost more difficult than the retrofit itself.

6. RECENT LEGISLATION

The high visibility of these bridges, the media attention, the high traffic volumes and congestion, and the resulting large revenue generation have made every aspect of these structures the subject of public debate and political activity. Recently, five laws were enacted that affect the operation and maintenance of these structures. The first, Regional Measure 1, was passed by the voters in the San Francisco Bay area. The second was proposed by the governor and passed by the statewide voters. The other three were passed by the State Legislature.

Regional Transportation Measure 1

Raised the toll on all 7 San Francisco Bay area bridges (to $1 from various rates) and earmarked the increased revenue for specific transportation projects including the widening of the San Mateo-Hayward Bridge, the construction of a new parallel Benicia Martinez Bridge and the replacement of the 1927 Carquinez Bridge.

Seismic Bond Measure

Approved a $2 billion state bond issue to retrofit public buildings and bridges statewide. Six hundred fifty million ($650 million) was designated for the seismic retrofit of the 9 toll bridges.

AB 192

Recognizing that the Seismic Bond Measure did not provide full funding for the retrofit of toll bridges, this bill transferred financial responsibility for normal maintenance of the San Diego-

FLOW CHART OF BRIDGE WORK RECOMMENDATIONS
STRUCTURES MANAGEMENT SYSTEM

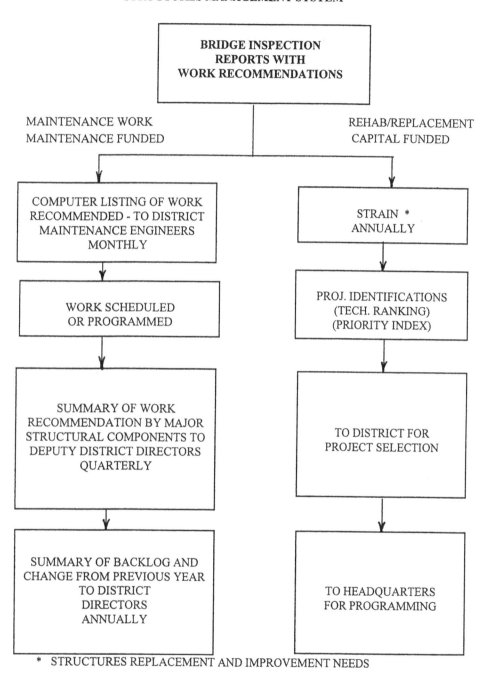

Figure 12. Flow Chart of Bridge Work Recommendations

ORGANIZATION CHART 2
TOLL BRIDGE OPERATION

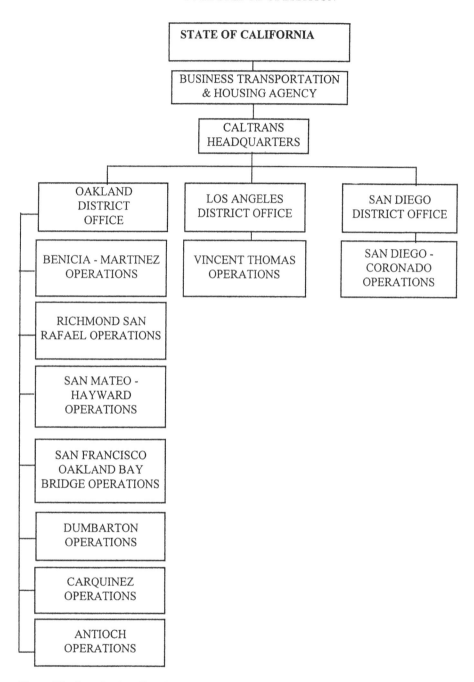

Figure 13. Organization Chart 2

ORGANIZATION CHART 3
TOLL BRIDGE MAINTENANCE

For Maintenance Performed by State Forces

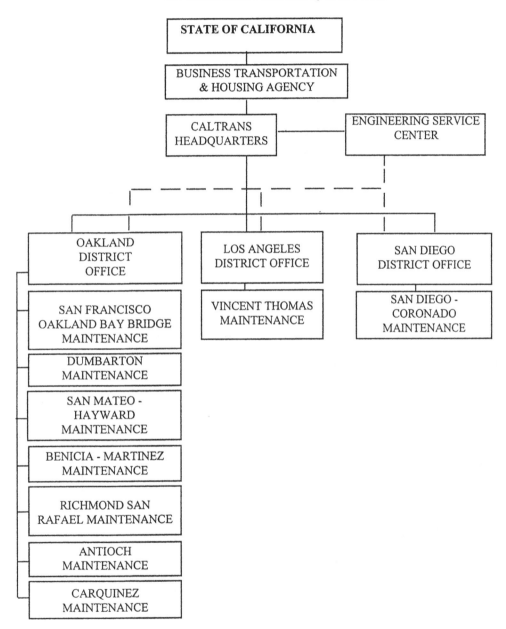

Figure 14.   Organization Chart 3

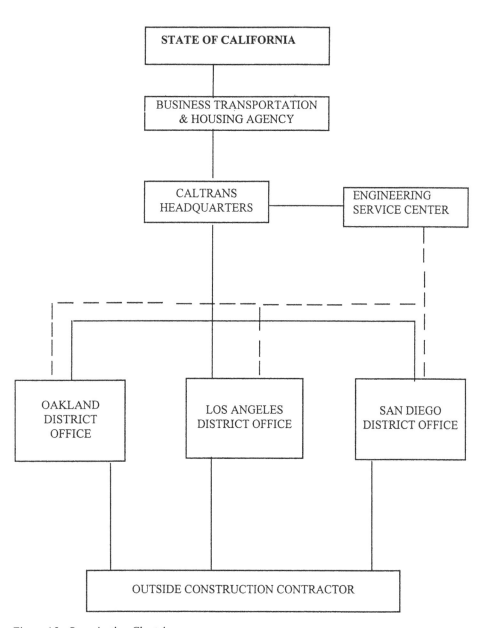

Figure 15. Organization Chart 4

ORGANIZATION CHART 5
TOLL BRIDGE REHABILITION

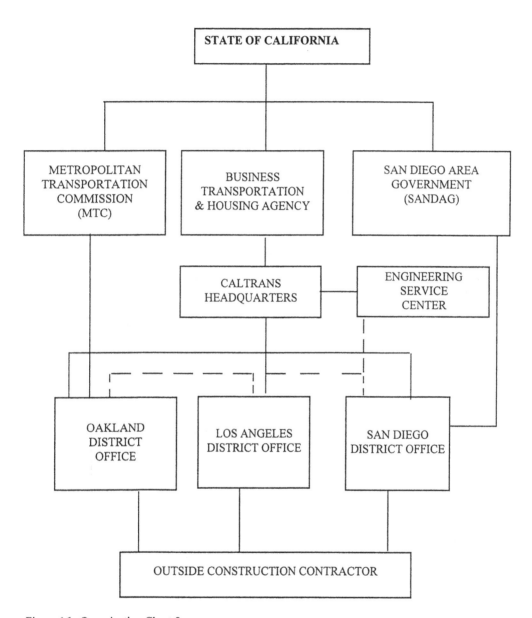

Figure 16. Organization Chart 5

ORGANIZATION CHART 6
TOLL BRIDGE SEISMIC RETROFIT

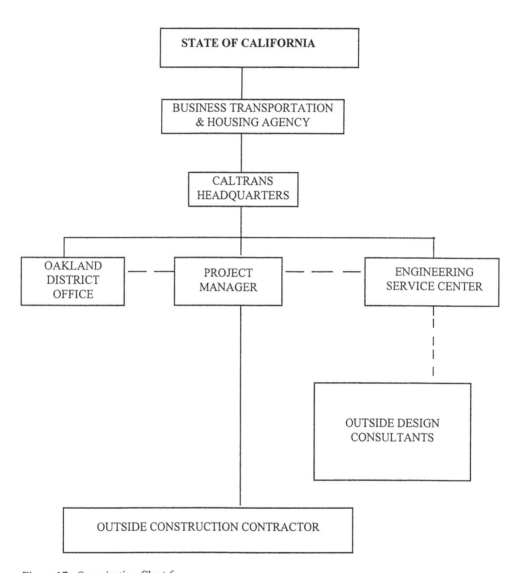

Figure 17. Organization Chart 6

Coronado Bridge back to the State Highway Account in return for a $33 million contribution from SANDAG toward the retrofit of the San Diego-Coronado Bay Bridge.

SB 47

Transferred the Excess Revenue Toll Fund for the 7 San Francisco Bay area toll bridges to the Metropolitan Transportation Commission (MTC), the local regional transportation agency in that area. Gave responsibility for programming and funding Regional Measure 1 projects and toll bridge rehabilitation projects to the MTC. Also allowed use of surplus funds for any local transportation needs.

SB 60

Recognized that previous legislation did not address the full scope and cost of the toll bridge retrofit program. Adopted an official cost estimate for each toll bridge retrofit project and transferred funds from various transportation accounts to fund these projects. Imposed a $1.00 seismic retrofit toll surcharge for all San Francisco Bay area toll bridges that took effect January 1, 1998. Allowed for a loan from the State Highway Account (SHA) to cover construction costs that will be paid off by the increased toll.

7. CURRENT FUNDING SOURCE

Operation - From bridge tolls, comes off the top before transfer to Regional Transportation Agencies.

Maintenance - From State Highway Account.

Rehabilitation - From Toll Accounts in the Regional Transportation Agencies except for the Vincent Thomas Bridge where the State still controls the Toll Revenue Fund Toll Revenue Fund.

Seismic Retrofit - A combination of General Revenue Bonds of the State of California and $33 million in SANDAG toll funds, $15 million from the Vincent Thomas Toll Revenue account and a loan from the State Highway account to be paid by a $1 toll surcharge on the San Francisco-Bay area toll bridges.

8. ORGANIZATION

Caltrans' general organization consists of a Headquarters Corporate Office, 12 District Offices and several central service centers. The Headquarters Office provides policy, programs, resources, oversight and guidance to the Districts and service centers. The District Offices are geographically distributed throughout the State and have responsibility for all Caltrans activities within their boundaries. The Engineering Service Center (ESC) acts as a program advisor to Headquarters for specialty items and as a consultant to the District Offices by providing specialty services.

Traditionally, Caltrans has been a functional organization, with line functions such as Project Development Right of Way, Construction, Traffic Operations and Maintenance, each having their own chain of command and responsibility for their portion of project delivery.

The Headquarters organization mirrors that of the Districts with each line function answering to the Directorate. The Engineering Service Center is also organized along functional lines and provides statewide Structure Design, Structure Maintenance and Investigations, Structure Construction, the Office Engineer, Structure Foundations, Materials Engineering Testing Services and several other statewide specialty functions. The general organization of the ESC is shown on Organizational Chart 1.

The primary functions of the ESC Office of Structure Maintenance and Investigations (OSM&I) are to periodically inspect the bridges and to recommend maintenance and rehabilitation work to the district. "Flow Chart of Bridge Work Recommendations" shows how the work is programmed. The bridge inspection report, prepared by an OSM&I bridge engineer, is the source document for programming and performing bridge maintenance and rehabilitation work. The bridge engineer also provides technical advice while the work is being performed. The actual maintenance work is usually done by the District maintenance crew. If the District maintenance crew cannot perform the work in a timely manner, a project to complete the work is programmed. An OSM&I design team or the Office of Structure Design then prepares the structure Plans, Specifications and Estimate (PS&E) for a construction or maintenance contract. The District prepares any roadway portion of the contract, including the traffic handling, and the ESC Office Engineer advertises the contract for bids and then awards the contract. Structure Construction provides technical support to the District in administrating the contract.

The next five organization charts depict how the various work is accomplished. Solid lines indicate a direct reporting or financing relationship, dashed lines indicate an advisory or consulting relationship. Many other Caltrans units support these activities, but only the major units have been shown for clarity.

Caltrans is in the process of converting from a traditional line function organization to a Project Manager organization form. The Project Manager would be responsible for delivery of a project from concept to the finalization of the construction contract. Under a "weak" Project Manager organization, line functions would report traditionally and provide services to the project manager. With a "strong" Project Manager organization, line staff reports directly to the Project Manager to provide functional services. Chart 6 shows the "strong" Project Manager organization for construction of the toll retrofit projects. Design of the toll retrofit projects was handled through our traditional functional organization structure, since design was started before we adopted the Project Manger concept.

9. CURRENT OPERATIONAL AND MAINTENANCE ACTIVITIES OF INTEREST

All tolls are collected in one direction only, with the opposite direction providing free travel. Tolls are usually collected from vehicles coming into a congested area, with free travel for those leaving the congested area which reduces traffic congestion.

Free tow truck service is provided to disabled vehicles on the most congested bridges to expedite traffic flow.

Permanent traveling scaffolds are being upgraded on the steel toll bridges.

Permanent in-house paint crews are assigned to the toll bridges in the severe environments. These bridges are continuously being painted.

10. FUTURE CHALLENGES AND EXPECTATIONS

The biggest challenge will come in the San Francisco Bay area where 5 of the 7 bridges will be undergoing extensive retrofitting at the same time. The retrofit strategy for the eastern end of the San Francisco-Oakland Bay Bridge and for the Carquinez West Bridge is complete replacement. The widening of the San Mateo-Hayward Bridge and the construction of a parallel Benicia-Martinez Bridge will also be concurrent. Getting enough qualified construction contractors and workers. and coordinating their effort with the on-going maintenance activities while maintaining traffic will be a Herculean task. The Oakland District Office has just created a new position, the equivalent of a traffic czar. He will have authority to stop or modify any construction or maintenance activity that in his opinion is unduly impacting public traffic.

The next challenge will be to work with the Regional Transportation Agencies to program and finance needed rehabilitation projects. These agencies have many competing transportation needs. Caltrans will have to convince them that keeping the bridges up to date and in top condition is in their best interest.

The final challenge I see will be to implement the Project Manager concept into our institutional culture. Our line managers are accustomed to having full responsibility for their portion of a project and are reluctant to give responsibility to a project manager. Future Project Managers must be identified and trained; a civil service career path should be established to make these positions attractive. Ultimate success of this concept requires that the Project Manager have authority to make resource decisions. This authority is now jealously guarded in Headquarters.

The future will bring automatic toll collection in some form, and most probably off-hour night work for all maintenance and construction that affects public traffic.

Political interest and activity has shaped our current organization model and that interest shows no sign of abating. Future political actions will continue to fuel organizational changes.

Operation and Maintenance of Large Infrastructure Projects, Vincentsen & Jensen (eds)
© 1998 Taylor & Francis, ISBN 90 5410 963 7

Author index

Printed and bound by CPI Group (UK) Ltd, Croydon, CR0 4YY

23/10/2024

01777679-0014